I0074986

RESEARCH SUB-CONTRACTOR

Models For The Decline Of The National Research And Innovation System

Prof. Mohammed Ahmad S. Al-Shamsi

Research Sub-Contractor

Management Models for The Decline of The National Research & Innovation System

Copyright © 2020 by Mohammed Al-Shamsi

All rights reserved. No Part of this book may be reproduced or used in any manner without written permission of the copyright owner except for the use of quotations in a book review.

First Edition

Published Date 2/2/2020

ISBN 978-1-7346287-0-8 (Printed Version)

ISBN 978-1-7346287-7-7 (Electronic Version)

TABLE OF CONTENTS

DEDICATION

*To the nation, who shall prosper by the
minds and arms of its people*

PRELUDE

Baghdad was the beacon of science and knowledge during the first centuries of the Abbasid era. Cordoba was a *shrine* for all people on earth who were eager to learn. Marab in the Sheba kingdom was the objective correlative of advanced architecture and a metropolis of that age. Likewise, Riyadh, or any other Arab city, could be so today if we so decide. Who erected these cities? We did! Our founding fathers did. All cities are parcels of land that surround us. Cities are not self-made; people built them and made them prosper through the accumulation of stock knowledge. Our predecessors who forged these great civilizations, put on a pedestal even by their enemies, proved themselves to be elites capable of developing and building. While abysmal ignorance shrouded the world, Arabs were enriching the accumulation of knowledge and exporting science to all peoples through many innovations in architecture, maritime navigation, medicine, and even warfare. Likewise, not only can we contribute today to the knowledge race to keep up with other nations, but we can also be at the forefront of this race. This is who we have always been, God willing.

The advancement and prosperity of nations and their stock of knowledge and ability to turn this knowledge into beneficial applications are interdependent. In the early 1980s, the nascent concept of national research and innovation systems advocated that the outputs of any nation be the outcome of all institutions in that nation, including organizational and management structures, financial and funding techniques, laws and legislation, and even the judiciary, which all constitute an integrated national system of research and innovation. Therefore, reforming the national research and innovation system requires comprehensive sectorial reforms; a fructification thereof would thus reflect positively on all sectors. Reforming and reconstructing the national research and innovation system is a nationwide investment to improve all aspects of life, be they construction, welfare, lifestyle, or culture. This would offer both combat preparedness when war approaches and leisurely pleasure when peace reigns.

I have been part of the national research and innovation system for nearly fifteen years, during which I assumed various management positions, accessed the data of thousands of research projects and their outputs, and had direct contact with national researchers and national institutions. Moreover, I compared and contrasted the national research and innovation systems to their foreign counterparts, where I worked for years. In light of this experience, I hereby analyze the fault lines in our national research and innovation system and propose some practical management solutions thereof.

I ask God Almighty to help me contribute to the rectification of the national research and innovation system before I leave and provide me with other honorable citizens to assist me. I write this to share my opinion with all stakeholders who may profit from my ideas and stand by my side in reforming our national research and innovation system to put them on the right track.

While writing the four parts and thirty sections of this book, I was keen to present these concepts in a readily accessible non-academic style for the reader so that it can appeal to a broad audience rather than a limited group of academics. Each section is a standalone article that has been briefly and narratively written in an easy-to-assimilate style.

God knows best the intention behind one's deeds!

The Author
30 January 2020

PART ONE

A System That Is NON-BENEFICIAL to a Nation and Wastes Its Resources

CHAPTER 1

Research Sub-Contractor: Concept Of Wasting The Wealth Of A Nation

A nations' development and the ability to attain significant progress depend on having an efficient research and development (R&D) system whose fundamental unit is the researcher.

Due to the petroleum economic revolution in Arabian Gulf Countries during the last 50 years, this region has a unique and peculiar management structure for their national R&D systems. This article adds value as a contribution to analyzing and determining a unique national system from within, which is unlike any other national system in the West or the East.

However, in most cases, distinguished researchers in our country have become *contractors* who receive funds from governmental research support sources only to channel them to a group of pioneering foreign researchers or distinctive academic entities abroad to obtain the highest quality research published in the most prestigious international scientific journals.

The quality of research receied by a contractor depends on the liquidity provided by the governmental entities. If the amount of funding is average, the contractor can frame formal research contracts with foreign research entities to conduct the required research.

If the amount of research funding is very high, then the contractor can sign a partnership contract or a cooperation agreement or establish a joint research center with the most outstanding international academic bodies. This choice depends on the size of the received funding. However, this type of center is a mere shell or cover for cooperation, with no single full-time researchers from either contracted parties; it only includes the names of some staff from both contracting parties to legitimize the transfer of funds to a foreign entity to conduct the research. The assumed gain for the contractor is that all scientific research published would feature the name of the contracting entity, as well as some of its researchers, thus testifying to the distinction of that research entity and its researcher, despite being a veneer distinction. This practice also assumes a high quality of research output, which is used as an indicator of the efficiency of those in charge of these research entities.

In light of this situation, one may wonder what is wrong with this practice. Is research collaboration and partnership not a universal requirement for scientific advancement? Do the diversity of researchers and their expertise not help push their ideas towards creativity and innovation? Yes, certainly. Then, why the criticism?

Only a veteran researcher from inside the system or a manager who has an overview of thousands of research projects funded by governmental sources and has observed their outcomes can probe the current practices and highlight the defects in the national R&D system.

We are writing to untangle this situation, so we may correct the path and save public funds by redirecting them to the true beneficiaries and the right channels. Moreover, this article will increase the awareness of researchers and guide them toward optimal research practices that abandon subcontracting. These entities and individuals are an integral part of the nation and the scientific research system therein. Perhaps it is a lack of insight, rather than bad faith, that is the cause of our current dilemma.

As a prelude to the argument, we will divide this study into simplified parts that will appeal to those who have had no leadership position in a research entity or are unfamiliar with the working mechanisms of R&D.

Argument 1:

In a specific framework, the national government's research funding comprises non-refundable grants aimed at developing the national research entities' tangible and intangible infrastructures, including human resources, equipment, and materials or for achieving a specific research target. In a general framework, such funding includes all direct and indirect R&D allocations from national government sources.

Ultimately, funds are allocated to the researchers at a national research entity, be they citizens or foreign residents. The same goal is achievable even if a foreign resident researcher is working for a national research entity. It is expected that such a researcher will inject their allocated financial resources into national laboratories and experimental fields and choose assisting research teams (e.g., technicians, lab chemists, engineers, laborers, and research assistants) to conduct the research project. If this is the case, there is no harm in funding a foreign resident researcher. However, we would rather fund national researchers because they are more sustainable assets for the country. In the case of an economic crisis, an expatriate could leave for better opportunities elsewhere, while national human assets could resist financial temptations because of other considerations, such as country, homeland, family, compatriots, etc., which could outweigh purely financial benefits. Thus, a wise investment in national assets is more beneficial and sustainable in the long term for the nation.

Argument 2:

It is possible to direct research grants to distinctive foreign individuals or entities by allowing them to apply for R&D grants directly and compete based on their scientific merit and proposal originality. In this case, part (if not most) of the funding will find its way out of the country to conduct the research, and we will thus achieve the desired scientific output at a lower cost while dispensing with research subcontractors. As part of research funding goes to the human resources of the subcontractors and their own research supporting team, this wastes part of the public funds on those, be they individuals or entities, whose only concern is bolstering their academic records.

However, in practice, we do not allow foreign researchers working for a foreign entity to apply for national research grants, even if they can conduct the desired research in the most professional way and within the shortest period, or if they have higher qualifications than local researchers and greater scientific publishing records.

Why do we then resort to this practice? This noble purpose is focused on directing funds to a national research entity or its researchers to establish, develop, and improve the infrastructure of the national research entity, without which we will never have a viable and accountable national research system capable of science and knowledge production and will remain forever dependent on others.

In other words, research subcontractors, whether individual or entity-based, have systematically demolished proper scientific research, undermining the national research and development system and wasting a substantial portion of public funds before transferring them to foreign entities who will carry out the research. Their receipt of research funding prevents its injection into the national research and development system, while transferring the same funding out of the country deprives qualified researchers and national research entities of their rights.

Hence, it is important and critical for the nation to correct this path by re-routing research funds to the right beneficiaries (i.e., national research entities) and abandoning research subcontracting.

Argument 3:

Can research subcontracting ever be acceptable? It can in two cases: First, when no research entity in the nation can develop or study an urgent research task (while not having the luxury of waiting for national capacity building) and we have not discerned its importance (though we remain accountable for past negligence and a lack of forecasting future needs of the country). In response to such urgency and due to the lack of ready local R&D entities, we must contract directly with a foreign research entity to do what we cannot do. In this case, we should not ascribe this research (and its scientific publication output) to any national research entity, lest we delude the public and the research entity itself that this is the optimal method and falsely polish the capabilities of the national research entities in the eyes of country leaders and in terms of indices and key performance indicators.

The second case is Ph.D. scholarships, for which we specify the research the candidate should undertake and send it to the most distinguished global research entity in this specific field study. This is a way to qualify national researchers for research that national research entities and their researchers cannot accomplish. The scholarship candidate would then transfer the knowledge he acquires to his country when he returns. Developing a knowledge decoding system is essential to preserve the values of the Ph.D scholarship program. Unfortunately, we send Ph.D. candidates only to obtain academic degrees based on their general field of study and sometimes based on the reputation of the school. No criteria have been developed yet to specify the absence of certain knowledge that the country will need in the future based on technological forecasting.

Although we have observed some pertinent individual cases, they are not systematically organized by research entity; this organization could be by done chance or by advance awareness of certain individuals.

Therefore, the individual cases we noted should be systematized based on the organizational structure of the research entities to serve the country, a matter in which we need to invest in the future.

Argument 4:

Notably, many research collaborations between national and foreign research entities are shallow. Indeed, cooperation requires that counterparts and their peers be equally (or close) to each other in their level of scientific expertise, so they can transfer expertise jointly to develop something to complement each other. Each cooperator has his own funding resources. Unfortunately, current widespread research collaborations in the nation are not achievements of the country, as some propagandists overstate; instead, they mainly comprise the transfer of funds from national research entities to an experienced foreign research entity to carry out specific research. In this way, this is a contractual financier–implementer connection. Simply put, it is a relationship between a funder and an implementer. This is not a healthy management practice for our national research system.

Argument 5:

If the current research subcontracting practice persists, our losses will be irreversible, draining the financial resources from our national laboratories and experimental fields and freezing the minds of our human resources because researchers will embrace the clear and easy route of fake excellence known as subcontracting.

CHAPTER 2

What is to be Done: Rebuild the National research and Innovation System

It is imperative to reform the national research and innovation system first if we plan to innovate and develop the latest technologies and high tech industries to compete with other nations and be at the forefront. Is our national research and innovation system invalid? Certainly. Were it valid, it would have produced the imported technologies used in most of our prominent major national industries.

To clarify and be more transparent, many major national industries, including the petrochemicals, oil, and date industries, among others, did not originate from national research, development, and innovation systems, and no national research entity has contributed to their establishment and progress. We imported such technologies and knowledge from foreign tech owners who manufactured the entire production lines for installation in our national factories abroad and have kept providing us with technicians and experts to maintain them. No national intellectual efforts or innovative contributions went into these technologies. When the first technologies became obsolete over time, we brought in newer companies and industries and imported the latest production lines for new technologies in many industries. Investors and company/factory owners have remained focused on the ever-changing developments in the east and west and have purchased these technologies to compete with their local and regional counterparts in the Arabian Gulf countries and the Middle East. This has been occurring without the assistance, development, or improvement of the national research, development, and innovation system. Investors and factory owners profoundly believe that foreign producers of technology alone can develop and improve this technology and invent the next generation thereof; thus, they have found no reason to count on the national research, development, and innovation system. Even if there are individual cases of national development, they do not prove that the national research system, as it currently is, is valid.

Realizing the aforementioned fact, those in charge of the research, development, and innovation (RDI) entities have adopted three routes.

Knowledge consists of a set of cumulative contributions that connect to generate value that may translate into improving an existing product or creating a new generation that supersedes its predecessor and creates a boom. Under this premise, national RDI entities believe that they could only contribute to such cumulative knowledge in one or more selected disciplines. Tens of billions spent on scientific research do not benefit the country and do not have a clear target. It is true that there relevant scientific contributions do exist, and we can provide outstanding performance indication figures for scientific publishing in this country in which we surpass or at least compete with our peers in the Middle East, but these performance indication figures are not the end goal. We seek the truth beyond these rather misleading performance indication figures.

The second route involves self-deception, misleading both the public and leadership. Some national RDI entities have been making various claims that look appealing, including the production of cars, planes, space shuttles, smart-phones, tablets, and many other items they never produced but rather imported under the legal cover of research cooperation or partnership.

However, this manipulation neither benefits the public nor turns a mirage into water.

Thirdly, there have been frequent intermittent, abortive, attempts by individual researchers or teams of researchers that fight for some time before they fade away due to the lack of or temporary support for their research entities. Individual efforts are always limited and ultimately discontinued. No institutional efforts have been made for national research entities to adapt continuous development processes to produce new science.

Then, what is to be done?

Before laying out the proposal, we must point out that the situation in the kingdom and the Arabian Gulf states is idiosyncratic and differs from that of Europe, America, and East Asia. Due to the economic booms accompanying oil discovery in this region of the Arabian Peninsula since the early 1970s, development accelerated, and the industry disbursements, mostly governmental, were huge. Neither the West not the East has ever faced such a dilemma, and they cannot provide us with any benchmarking situations to provide us with a ready-made solution. These unique and peculiar regional characteristics render it necessary to determine a unique solution from those who have assimilated the system from inside rather than from foreign consultants who may find it difficult to understand these interior complexities and who may think that solving this dilemma would make us competitors in the future, which we can surely become, given enough time.

First, we need to remove the privileges and benefits associated with scientific publishing through the research subcontracting system for individual researchers or for national research entities by developing a national index that screens works in terms of their actual contributions from national research entities ' researchers. We already developed such an index in one of my previous works, but, unfortunately, it is not in use and has not yet been adapted. In this way, we can measure individual researchers' contributions to the nation and determine how original and authentic their work is. Although we do not claim complete that this index is fully precise, it would help enlighten the leaders of national research entities on the real quality of their work, separate from misleading allegations. If we can uncover the truth, wrongful practices will stop, and the country will steadily progress.

Second, we need to issue a regulation that criminalizes fictitious research and research subcontracting, which outlines the relevant categories and related penalties. Tolerating such practices compromises the reputation of the country and the output of our national RDI system, thus undermining its output, casting doubts on its potential, reducing its private sector support, and even pressuring the government to marginalize it under the pretext of having no return.

Third, we need to have strict management of national research, development, and innovation (RDI) and never allow unqualified researchers to assume any positions in the RDI system. This should include strict merit-based employment rules distinct from favoritism, the creation of high-standard scientific councils to promote those who are deserving, and the introduction of a non-lax scholarship system that allows research entities to impose maximum penalties, including dismissal and the refunding of scholarship expenses. Real firmness is good for the progress of the National RDI System.

Fourth, we need to anticipate the required technologies and identify specific scientific fields to establish their respective infrastructures and introduce structured long-term funds to mature and produce the envisaged technologies and industries. For example, we can certainly produce jet aircraft, smart cars, submarines, space shuttles, intercontinental ballistic missiles, and weapons through the national RDI system, but we cannot do so in one, two, or even five years. Establishing the infrastructure for something new in place requires at least one to two decades. The researchers or those in charge of research entities who seek to polish their images must stop their misleading practices and respect the minds of the country's leaders, the public, and their colleagues. We can accomplish these goals, but we need a long-term commitment of no less than a decade if we are to start a new industry that does not already exist. We must ensure that every new leader of an R&D entity adheres to a long-term strategic plan unless there are strong, urgent, justifications for any changes or replacements. It will be a disaster if every new leader redefines research priorities according to their whims, regardless of the progress of their predecessors; if this persists, we will never be distinctive.

CHAPTER 3

Repositioning A Nation: The Prosperity of Education, the Decline of Research, and the Importing Industry

The educational renaissance in the Arabian Gulf Countries, including Saudi Arabia, has progressed over the last 60 years like no other region in the world. Currently, Saudi Arabia has about two million holders of university degrees and more than twenty thousand Ph.D. holders, compared to only hundreds before the beginning of this renaissance.

Although Saudi Arabia and the other Gulf states spend hundreds of billions annually on education, the national RDI system has not produced any distinctive national industry. We have imported many existing (neither few in number nor superior) industries, namely technology, production lines, and know-how, with no touch of our own in terms of development or improvement. We have been consistently importing and receiving new technologies from their foreign owners–exporters, with no innovation and development of our own. In other words, the national RDI system has help no place in these major national industries (except for a few insignificant cases). It is even more sobering to know that we have not yet started our journey towards technological independence; we are still dependent on foreigners, be they individuals, entities, or states.

Our large number of university graduates have not improved the national RDI system. Where is the gap then? The issue is that these university graduates can barely operate these imported technologies and machines on the production lines and in the oil fields (where a substantial number of foreign human resources work due to their advanced skills and knowledge). In other words, this has been a quantitative rather than a qualitative increase. The national RDI system has not actually benefited qualitatively because, when these graduates join the research and support jobs, they do not develop reliable research systems nor do they develop industries, technologies, products, and know-how, or at least contribute to the improvement thereof. We cannot deny that research entities and their researchers may have made contributions, but these contributions have not yet produced the technology we use in our production lines for weapons and ammunition or even daily goods and services. The national RDI system did not even allow us to start production on the production lines of ammunition manufactured and installed by others, which did not develop until we purchased new technology and updated the production lines through the foreign owners of the original technology. This also applies to pharmaceuticals and the drug industry, where production lines manufacture products whose intellectual property has expired. The national RDI system has not had any impact in this regard in terms of development or improvement, even after three decades of technology transfer. This intertwined triangular combination (education, research and innovation, and industry) has not yet yielded the desired outcome for our nation.

What should we do to link national industry to the national RDI system?

First to start, we have to separate education from RDI to avoid any confusion in establishing their respective priorities. Then, we need to develop indexes specific to each sector so that leadership can distinguish and evaluate the performance of each industry separately, if necessary. It is also noteworthy that research in Saudi Arabia does not receive a fraction of the care afforded to education.

Second industry should stop bragging that the imported production lines in factories are national products, while providing facilities for the use of products created based on our national RDI systems.

Third as for investors steering the national industry according to the expected returns on profits, these investors will choose either foreign RDI systems or the national RDI system. Technology feature the same trend. However, the risk is high for investors to trust an immature national RDI system that has produced nothing compared to the mature, readily tested, technologies produced by foreign research, development, and innovation systems. Therefore, urging investors (those who steers the national industry) to deal with the national RDI system without a package of government incentives provided by the Ministry of Commerce and Investment and others will be futile. This approach is tenable given that many major national companies are state-owned or state-financed, and it will thus be easy to direct them in this way for the long-term public good—i.e., the future benefits that the country requires, rather than investors' immediate benefits.

Fourth   the national RDI system, which consists of research entities, regulations, legislation, and human assets, is the cornerstone of the nation's renaissance and without which we cannot excel in the industry. In this context, education for education's sake (i.e., education that does not feed the national RDI system), is useless for industry. In my previous article, I explained the way to rectify the national RDI system. I reiterate here that improving, mentoring, and refining the management of these entities is the optimal route to rectify the entire national RDI system.

Fifth   patience, perseverance, and farsightedness are needed to race against and overtake other nations. Any goals or visions we create today will not materialize in a year or two. When we aspire and plan, we must realize that the realization of our hopes requires unrelenting effort over time. Without this understanding, we will achieve nothing.

CHAPTER 4

A Nation's Ambition: The Outcomes of Foreign Research Systems Ensure The Prosperity of National Industry

If this issue looks complicated, we can deconstruct it here. How does a link exist between the national RDI system and the major national industries? If we can understand the dynamics of major industries, we can easily determine the dynamics of small and emerging ones. Existing major national industries generate economic value for the country, but we import all technology therein. Under normal circumstances, if we do not interfere (which is the case now), these major national industries remain dependent on the products of foreign RDI systems in the form of development and innovation packages shipped into the country. In this way, the major national industries can develop and import in the latest innovations without the need for the national RDI system.

What should we do to link industry to the national research system?

Let us differentiate between creating a new industry that does not currently exist and developing an existing one. Let us start with the latter type of development, because it is somewhat easier. The existing industries in the country are either significant and broad-based or nascent with no significant economic return, at least for the time being. Current economic indicators can help us determine the types of industry where the income constitutes most of the national GDP. We need then to divide every major industry into segments, which represent the industry's branches, and analyze them according to their respective contributions to the GDP and determine the most important segments accordingly.

The second category of significant industries may be small scale, have low economic returns, be under construction, or include industries that have not entered the country yet. Using economic indicators, we cannot distinguish critical industries that have not yet matured, but these industries may be more important to the country's future. Therefore, we must be careful about forecasting the unknown.

After identifying the major and future critical industries present in the country, we then form diverse research teams (dedicated to this research work) to study the production lines in the relevant factories. There should be a research team for each production line or a group of identical production lines in each industry to study and identify defects, shortcomings, aspirations of the operators, and consumers' ambitions about products and services. These research teams will then develop a priority list for the developments required in this industry, followed by devising solutions for each challenge, testing them in laboratories, and commissioning them under different conditions. The accumulated development results would be integrated into a newly developed production line. The numbers of each production line would eventually reach tens and hundreds, rendering each line unique and unparalleled due to its added features. In this way, the developed production line would be attributable to the national RDI system. Over time, we would be able to make unique products. We could apply the same process to commercial activities, such as treatment methods in hospitals, on-site remediation mechanisms, etc. We can identify the shortcomings and defects of such services just as we do with products, along with the relevant consumer expectations; the development of service would follow suit.

One may inquire how to manufacture the giant equipment destined for production lines (commissioned in laboratories and research experimental fields). This process is not easy, but it is possible. Those who design experimental production lines in the laboratory can design the larger equipment for a factory. This process begins with designs and then an application of those designs on the ground. Moreover, we can obtain the designs of original equipment (before development) through reverse engineering and then add adjustments. If these adjustments are in the hundreds, then the performance and efficiency of the new product will differ, which is the whole point of development and innovation. At that juncture, the investors steering the industry will sense the importance of funding and cooperation with the national RDI system. They will compete in supporting distinctive research entities and researchers. Over time, the country's economic returns from the national RDI system's outputs would multiply.

The creation of a new industry involves an industry that is either new to the country but exists somewhere (such as nuclear power plants, and space rockets, and submarine factories), or an industry that is completely novel and not present in the country or elsewhere; the latter is what we aspire to and need to make happen. This is the reason for writing these articles. We should not only copy foreign RDI products but compete with them.

It is easy to bring in a new industry that is available in another country but new to us. This process requires government investment, facilities, and pledges to persuade investors to take on something riskier than traditional pursuits. The transfer of a new industry can be done through purchasing a production line, even an old one, from a technology owner in a foreign country. Then, we can assign the task to the abovementioned research teams. If it is not possible to purchase a ready-made production line, even an old one (as with weapons, space, the nuclear industry, etc.), we can start with the product itself, such as an airplane or a weapon, which research teams can analyze and develop approximate designs for using modern reverse engineering tools and devices. Then, we can manufacture prototypes in workshops and commission them in experimental field tests and laboratories to identify a series of developments after identifying the product's deficiencies and gaps. After overcoming all obstacles and improving the product, we can create a real production line and then complete the abovementioned development processes.

Transferring a new non-existing service is more difficult but possible by sending qualified researchers on Ph.D. scholarships or post-doctoral programs into the relevant field to acquire the necessary know-how. Upon their return, we can establish service companies where they these scholars will work as consultants to transfer their know-how to new companies. These are more favorable scholarship programs than unsystematic scholarships that provide graduates whom the domestic labor market does not need, resulting in a non-stop unemployment crisis.

Other techniques do not originate from the national RDI system, including attracting companies to provide services in the country and offering all facilities for five years, provided that 50%–75% of its staff are nationals who will then continue the services after the foreigners leave. However, we need to fulfill our commitments towards foreigners to maintain our reputation and encourage others to come if we need them in the future. These facilities should be limited to three to ten years of operation due to many considerations. For example, when the foreigner is the only researcher versed in the technology (service) we need to acquire, he may control any competition that may arise after the established period.

We devote a separate chapter to the *modus operandi* of the national RDI system in the production of new non-existing industries around the world.

CHAPTER 5

Universities Are Not Product Factories

The public and some leaders of national research entities confuse the university (or any research entity, be it a center, an independent research institute, or a subsidiary of the university) with a factory or company. The fact is that a university is not a factory that produces goods and commodities for markets to gain profit. The same concept applies to service companies.

The university is a platform for exchanging ideas and an interaction medium for its staff, students, and society. A university provides an intangible product, knowledge, which should be its optimal contribution, and its main key performance indicator should knowledge accumulation. This determines a university's impact, given that it has no commercialized commodity. None should consider the university's output as similar to that of a factory or company.

In his book, *The Uses of the University,* published in 1966, Professor Clark Kerr, former President of the University of California, explains the role of universities in imparting knowledge. He notes that universities provide a great service to humankind (i.e., contributing to the knowledge record), which may be used later (not by the university) to create marketable products and commodities.

To link these concepts with practice, the first startups founded in Silicon Valley were created by Stanford University professors and students due to the policies of Professor Fredrick Terman, the school's rector at the time. Terman encouraged students with ideas for new technology projects to leave the university, and he helped them set up companies in Silicon Valley. He insisted that the university should remain a sanctuary for generating ideas and innovations, rather than metamorphosing it into a manufacturer of goods.

Where is the border between the university and a factory or company? We need to answer this very critical question for all presidents and leaders of research entities. The university must know its limits and never metamorphose into a factory, even if a commodity is the outcome of its innovation, or if a market analysis indicates that the university's profits would exceed its expenses and constitute an additional income. Each stakeholder in the national research and innovation system must undertake its *role* professionally. Companies are an integral part of the national research and innovation system that receive ideas and innovations from universities and research centers and convert them into production lines in their factories.

A share of intellectual property royalties will guarantee a university's rights and efforts as it continues generating and injecting new ideas into existing companies and startups. This would be a win–win situation in terms of economic returns and benefits. In light of this, we should be wary of calls to transform universities into companies that produce goods and commodities and provide technical services to consumers. This would be a deviation from the true objective of higher education and a waste of public resources.

In this context, the ideal model we should apply to universities and research centers is to keep faculty staff and researchers focused on generating ideas and theories and urge students to establish companies to apply them. This is an idealistic proposition, which does not purport to prohibit researchers from creating technology startups. However, ideally speaking, we need researchers to continue generating ideas and presenting innovations, not to move into industry and abandon their noble profession.

CHAPTER 6

Research Entities Fix Production lines and Sell Products

We have noted that some national research entities have turned some of their spaces into commodity factories. Let us highlight the flaws in this uncommon practice. First, the goal of any research entity is not to brag about its ability to produce and thus delude the public that it has a high production capacity and tangible goods in the market. Unfortunately, such research entities misunderstand their role and expected contributions to the nation and the community.

This practice involves a series of mistakes. First, these goods and production lines are neither an innovation of the research entity nor the fruits of the ideas of its own staff. Thus, they do not embody any knowledge created by the university; this entire endeavor is mere imitation. Second, engineering companies and the construction companies installed them feature no contributions from the faculty of the research entity. Third, the operation of these production lines (not to mention the capital needed to construct them) deplete the public funds allocated to the research entity for use in research to advance science and develop ideas and innovations. Fourth, these factories and production lines placed in the research entity remain the same, even after the lapse of more than a decade, with no significant contributions by the research entity's faculty in terms of the development of the production line or the improvement of the commodities they produce. Fifth, this research entity has not exerted any efforts to develop and improve the products or commodities in these production lines. A lack of successes in development or innovation is sad, but not ever trying is miserable.

These research entities may build production lines, but when it comes to selling goods, they should engage a marketing agency. If they absolutely must, production lines should be of a small scale for simulation only and represent the brainchild of the research entity's faculty; if they are not, then they should be for developing and improving production lines in a microenvironment. If the research entity is unable to develop these commodities, it should at least support a series of studies aimed at improving the commodity or changing the production lines.

If the research entity does not engage in any of these measures, the process of establishing and operating the factory or production line is misguided, misplaced, a waste of the public funds, and an indication that the person(s) in charge of the research entity is/are not wise.

CHAPTER 7

Increasing The Expenditure Of R&D Is Not The Optimum Solution

A report published by the Ministry of Education indicated that Saudi Arabia has spent about 24 billion Saudi Rials annually (approximately 17 billion Saudi Rials from the government) since 2012 by adopting the Frascati methodology to calculate R&D spending, and that spending was 12 billion Saudi Rials since 2010. In other words, Saudi Arabia spent more than 100 billion Saudi Rials from 2010 to 2020 on its research and innovation system. Why then do not we see an industry or technology produced by this national research & innovation system? Where is the problem?

Let us shed some light on the RI annual spending of research and innovation systems in developing countries. India spends more than 200 billion Saudi Rials annually, Brazil more than 100 billion Saudi Rials, and Israel and

Mexico more than 40 billion Saudi Rials. These are all developing countries that possess industries resulting from their national R&D.

Accordingly, the easiest solution here may be to ask the government and the private sector to increase their R&D spending close to that of other developing countries. Nevertheless, this is not the optimum solution. It is true that increased spending means additional opportunities to create and develop academic infrastructure, but sector governance requires a complete overhaul, as Joseph Schumpeter explained in his theory of Creative Destruction[1]. We also need to ensure that the disbursements to national research and innovation entities actually reach their laboratories and testing fields and do not leak abroad, as we explained in previous chapters.

To reposition national research and innovation governance, we should implement the following major reforms:

- Prevent the leakage of funds pumped into the national research and innovation system outside the country.

- Give more attention to management and leadership in the national research and innovation

---

1 "The process of industrial mutation that incessantly revolutionizes the economic structure from within, incessantly destroying the old one, incessantly creating a new one."

system by providing intensive programs to train researchers before they take management and leadership positions and throughout their management careers. Add to this establishing a national academy to qualify the leaders of national research entities, just like the Command and General Staff College for the military and the Diplomatic Institute for the diplomatic corps.

- Revisit and reposition performance-based incentives for researchers.

- Provide a package of benefits and facilities for investors to direct their funds to the national research and innovation system.

- Develop an index to measure the quality of national research entities and allow them to nurture the markets with new products and services. This index is important for distinguishing research contracting entities and true research conducting entities.

- Develop an index to measure each individual researcher's contribution to the national research and innovation system to identify the right destinations for financial resources.

- Establish a national council on the credibility of research entities and their researchers' outputs to review the products claimed by the academic entities and researchers beyond scientific research and publishing, whether on the market or forthcoming.

- Establish a national authority for scientific credibility to hold the researchers breaching this credibility accountable.

- Issue strict regulations and sanctions that enable the relevant government agencies to deal with violations of funding leakage, a lack of credibility, and breaches of scientific trust, with penalties up to dismissal, financial fines, imprisonment, suspension, and a refund of misused funds.

- Withdraw the powers for scientific promotions from academic entities and create a national unified scientific council that receives requests electronically and impartially without direct contact with the applicants. Academic promotions constitute the greatest incentive for researchers to engage in scientific publishing and innovation. Therefore, reforming the promotion system is good for the nation.

- Develop hiring regulations in the national research and innovation system to ensure minimum standards and remove exceptions. Current hiring regulations for academic staff are not strict enough to select the best researchers based on pure merit.

- Reform the research funding system and issue regulations that enable academic entities and researchers to receive and spend money from investors and companies practically and smoothly.

- Issue a legislation that allows researchers to establish startups without bypassing the system or losing their jobs.

In the fourth part of this book, we will provide details for some possible proposals.

.

PART TWO | Innovation Generates a Nation's Wealth

.

Sustained and Disruptive Innovation: Saying Goodbye to the Future by Dispensing With the National Research System

In 1997, Professor Clayton Christensen, in his book the Innovator's Dilemma, provided an appropriate explanation for the separation of sustained and disruptive innovations. We will begin by introducing the concepts he referred to and then link them to the national research and innovation system and how the second type of innovation is rarely produced outside the research and innovation system. At the end of the article, we will highlight the inherent risk in dispensing with the national research an innovation system and relying on products from foreign research and innovation systems, namely our inability to create the next generation of specific technologies that will replace the current ones. In this way, we bid farewell to our future because the current technologies used in the national industries are a product of foreign research and innovation systems, and future technologies will also be from them, unless we correct our path and rely on our national research and innovation system.

Sustained Innovation is one the expected outcomes of innovations to develop existing products and services. A specialist can identify shortcomings in a current product or service or the benefits required to meet the needs of the user or a beneficiary. This is the type of innovation targeted by companies and industries, and this is what investors pay for in research and development, whether by allocating money to in-house developers/centers or through cooperation with the academic sector. In general, the expenses of a company in the academic sector have three destinations. The first is allocation to research chairs for specific needs, which leaves the academic body free to engage in research and provide the outcomes of these works to a company as a beneficiary, such as a research chair developing a drug to treat a specific disease. The second involves a work contract with an academic entity (or more than one entity in one contract) to improve a specific service or product. The third destination is directly to a distinguished researcher or research team in the field that has well known scientific publications that enable the company to reach them and assign a task. The third destination is the lowest in terms of costs and the fastest in terms of results, the corporate governance of output, and easier business processes for the company. However, this practice is often challenged by some research entities whose work is not completely under the funding's umbrella, resulting in a loss or neglect of the academic entity's right to intellectual property or the expected proceeds from the envisaged product.

Disruptive Innovation refers to that which the specialists and experts in the company or industry cannot know or determine their desire to attain; this emerges in isolation from existing industry. This type of innovation is not attainable through the desire of a company or industry, in-house developers, or in-house research centers within the company. On the other hand, neither a research chair nor a research contract with a research entity can be sufficient for this purpose, nor can a prominent researcher solve this "unknown problem". This type of innovation precipitates a boom that supersedes an existing industry. For example, cellular phone technologies have virtually eliminated fixed-line phones and have acquired the market share of landline phones. The development of personal computers has likewise replaced supercomputers and has acquired their place in the market. This also applies to giant mechanical mills whose market share faded in favor of the smallest mills that are the most practical, cheap, and easy to use and maintain. We return now to an example of an Arab industry that once occupied a world-renowned position: the natural pearl industry in the ports of the Arabian Gulf coast. In the past, pearls were a source of income and reliable industry throughout the world. In the 1930s, the Japanese managed to produce artificial pearls, which suddenly swept the market due to their cheap price and the inability to differentiate their quality from natural pearls, except by experts. Artificial pearls took over and replaced natural pearls in the market, resulting in the depression of some cities of the Arabian Gulf coast. No one could have predicted this would happen, but it resulted in the loss of many jobs and diminished the sales of stores and ships.

For further clarification, those in charge of the current industry will be surprised by the upcoming generation of technology that will replace the current industry and take its share in the market. The timing of this development is unknown and unpredictable. It will happen suddenly and fall catastrophically on the people in the present industry. No innovations or developments in the current industry will prevent this. Investors should keep in mind the products that are currently under development in the research laboratories and the innovation entities that are not within their industry. We know that this is a difficult task, if not nearly impossible, but this is what happened in the past and what will recur in the future. Therefore, investing in the national research and innovation system is the best solution to produce next generation technologies that will capture the market in place of current goods and services.

Investors control companies and industries. However, an investor does not see a difference if the product emerges from a national or a foreign research and innovation system, as he will obtain the latest technology from abroad if it is not produced from within. Herein lies the role of the state in developing and improving this sector and their role in creating a package of benefits for investors to transfer new technology from the laboratories of national research and innovation entities. Without state interference, this sector will not grow.

The only catalyst for domestic investors without state interference is a ban on the transfer of some technologies from abroad and the restrictions that foreign countries place on this transfer, such as the technologies for nuclear weapons, among others. If these weapons were a product of the research and innovation system in the country, no one would dare prevent it or place restrictions on it because, in this case, it would be our knowledge and product. In that case, we could export the fruits of our labor and sweep foreign markets with unparalleled products and services. Moreover, the infrastructure may not be ready to emulate these products in foreign countries for a few years or even a few decades, depending on the complexity of the technology we develop.

The other catalyst without state intervention is the patriotism of investors, which is the subject of Chapter 16 in this book.

Chapter 9

Germany, Japan, and We:
The Fourth Industrial Revolution
and the Fifth Social Revolution

The creation of research entities alone is not sufficient to achieve innovation, develop economic resources, and feed local investments from the results of these innovations and human knowledge development. First, we need to establish financial and funding institutions on the one hand and legal institutions on the other hand. However, the existence of these three categories of institutions is not enough; it is only a starting point. Without developing cooperation and synergy and enabling the flow of business between the three categories of institutions, the national innovation system will not operate.

The national research and innovation system requires the synergy between financial, academic, and legal institutions to serve the commercial and industrial institutions, unless we want scientific research outputs to remain locked in drawers. It is not possible for one ministry or body in the state to mandate synergy or facilitate the flow of business between the three categories of institutions. This process requires synergy between at least four ministries and a group of support bodies.

If we take Japan's new blueprint for a super-smart society (Society 5.0,) developed as an advanced and updated version of Germany's "Fourth Industrial Revolution" (Industry 4.0) and look at the communication structure, we find a mechanism to link several sectors in each field to a program chief directly linked to the Prime Minister. This is done to facilitate communication, solve obstacles that arise, and enable rapid escalation when communication is not possible or performance is delayed.

The role of the government is essential in creating an ecosystem that enables innovation and facilitates the flow of innovation to factories and then to markets by being the only entity that can link, activate, and enable the work of the national research and innovation system. The ability to enact and enforce related policies is only part of the government's abilities, in addition to the support, guidance, and mentoring it can provide to the private sector.

Accordingly, to activate the national research and innovation system, we recommend linking all the entities concerned to a high-level coordinating council that reports directly to the Prime Minister. This council would offer a high-level and comprehensive view to link the relevant sectors in the state, facilitate all the ways to enable effective communication, facilitate the smooth flow of business between different sectors, re-formulate and draw up policies to suit these linkages, and resolve conflicts and duplicity among the relevant powers and tasks. Under this Supreme Coordination Council, an executive body would be responsible for fieldwork to map the flow of innovation outcomes, from the creation of ideas to the market, as innovative products that yield an economic return and wealth for the country.

Chapter 10

Technology Valleys and Parks:
A Beginning below expectations

In her book published in 2019 "The Code: Silicon Valley and the Remaking of America", Margaret O'Mara noted that technology valleys are combustion chambers in the economy of the United States of America that have created companies that have become a model for the country's largest technology companies. This book indicates that the value of the five companies that came out of the American technology valleys (Apple, Amazon, Facebook, Google, and Microsoft) is equivalent to the entire UK economy.

During the past decade and a half, the awareness of decision-makers in universities and various academic sectors has increased, as reflected in the establishment of technology parks and valleys. Universities have established private companies to manage these parks and valleys, including Dhahran Techno Valley, Riyadh Techno Valley, Makkah Techno Valley, Taif Techno Valley, and others. Legislation has enabled academic bodies to operate these valleys freely, with powers granted in financial and legal affairs, which is a leap forward. During the fifteen years since the start of this technical revolution, dozens of companies have had their headquarters in these valleys.

However, this launch was not in the right direction. To make this clearer, let us look at the origin of the majority of the companies in these parks and valleys. These areas were facilities to host major international and local companies, not small startups. After pumping billions of Saudi Rials into these valleys, we find only a few dozen of national startups.

These techno valleys and parks are places designated for startups and entrepreneurs, which provide a set of facilities and privileges for beginners in establishing non-traditional companies using their business model, with a high risk of invested capital and a low chance of survival. The chance of failure for companies with radical ideas is much higher than the chances of their success. Therefore, these companies need special care and a special package of facilities, the most important of which is capital, a workplace, and some legal and technical advice to enter the market.

Accordingly, we need to revisit the management model used for these techno parks and valleys to correctly direct efforts to serve promising startups that need a space and not to allocate these spaces in techno parks to serve major global or local companies.

The ability of a techno parks or valley to host major international or national companies that have been in market for decades is not an achievement, nor does it offer benefits for the country; it is also a waste of wealth. The allocation of an office or building in a techno park or valley in the country to a company with capital equivalent to the economy of countries will not lead to the success of a company that is already successful and already present in the local market. Giving such companies a building will not open an outlet for them on the local market, allowing us to enjoy new products and services that do not exist. Of course, giant companies do not mind opening an office in technical valleys after they are provided all the facilities, benefits, and exemptions for no less than five years or more. What some consider an achievement is really a waste of national wealth and not directed to the actual beneficiaries, namely the ambitious citizens who aspire to obtain a tiny fraction of the facilities and benefits provided to those who do not need them.

Again, we are not in support of increasing restrictions and controls on academic entities to limit their powers and not standardizing their procedures in relation to innovation. Giving these entities a certain capacity and freedom to act is required; different approaches and management and executive methods may distinguish one of these entities and provide a better model for others or for new techno oases and valleys. Establishing a body or council to organize the affairs of these oases may not be a desirable measure because of its restrictions on creativity. However, we need to empower an existing authority to prevent directing resources in the wrong way.

Our ambition for these technical parks is great, and our hopes are greater. We may be proud, someday, even after a decade, of startups that originated in these parks and then covered not only our nation but also Asia, Africa, Europe, and the Americas with their services and products. Then, we will all have a patriotic model whose achievements we can be proud of.

Chapter 11

Business Incubators vs. Innovation Incubators

The first business incubator was established in New York in 1959 in the name of the Industrial Center of Batavia, which arose because of an economic collapse in that region. The companies that rented a large building area left after incurring losses, and no large company wanted to rent these buildings. Therefore, an idea arose to rent these office spaces at nominal prices to small companies as an incentive while providing them with some legal and technical support and common facilities, such as meeting rooms, coffee rooms, general reception, and toilets. The proper concept of business incubators did not emerge until the 1970s. The International Business Innovation Association notes that there are more than 7000 incubators and business accelerators around the world today. Innovation incubators emerged later.

Business incubators and accelerators contribute to the creation or support of newly created companies with unfamiliar or uncommon business models. Innovation incubators contribute to the creation and support of newly established startups that have an innovative product or service with intellectual property rights that enables the founder of the emerging company to protect this product. Innovation incubators produce startups with intellectual property rights, and this is what distinguishes them from business incubators. Intellectual property rights protect a start-up owner against the replication of his work when it becomes successful in the market. However, startups that do not have intellectual property rights produced by business incubators and accelerators face difficulties after their success in terms of fierce competition. If they succeed, others would like to replicate the same business model, which has not been kept secret or duly protected.

Companies with intellectual property rights can face all competition that may arise, except by producing another innovative business model that has intellectual property rights and better market advantages. Thus, competition becomes based on the innovation and introduction of new goods and products into the market and not on replicating the same work with slight improvements or a reduction in price. This competition has no value for the country, and it may become harmful to the economy when competition is limited to lowering prices. Harvard professor Michael Porter, in his works, elaborated this concept. Competition for innovation, however, contributes to the prosperity of the nation as a whole. When these innovative companies expand, and if there is good management of their intellectual property, the products and services of such companies can enter other countries in the East or West without competitors because the right to protection in those countries protects such products and services from competition and imitation. In this case, the wealth from other nations will flow to the capitals of our country.

For comparison, since 2006, dozens of business incubators and accelerators have been established in Saudi Arabia, both in the private and government sphere, with great diversity in their activities, and in fields from information technology to biotechnology. These incubators have contributed to the establishment of many startups. It is true that many of these incubators have disappeared, but the worth of some have exceeded one billion Saudi Rials, which indicates a success. Our ambition is still greater, but the beginning of this process has been successful, and the models we see are promising. As for innovation, there are no dedicated incubators, although some startups have been established with intellectual property rights that protect them when they are successful or expanding.

In short, we need innovation incubators. This does not mean reducing the value of business incubators and accelerators. They still have an added value to the national economy. However, these incubators need a further increase in venture capital to attract and simplify their procedures to suit the general public and increase their numbers to include all cities and fields

Chapter 12

Technology incubators have a higher economic return than business incubators

Specialized technology incubators have a more specialized advantage than general business incubators, which do not necessarily aim at creating and supporting companies based on innovation and do not require intellectual property rights. These incubators focus on specific technologies, such as nanotechnology, pharmaceutical techniques, environmental technology, and the climate. What distinguishes these incubators is the addition of laboratories or experimental fields and the provision of some specialized consultants. Establishing this type of field of incubator costs more due to the need to have experts not only for legal and business consultations but also based on the specialty of the incubator. The incubation period needs to be longer depending on the complexity of the specialty and the minimum technical skills required. Receiving incubation requests in these incubators for non-specialists who do not hold degrees in this field is arduous and painstaking and turns them into training centers instead of incubators to establish and support startups. Accordingly, the segment of the beneficiary is always less than that of public business incubators.

However, the added economic value of these companies when they enter the market is always higher, even if they do not have an innovative product or service protected by intellectual property rights. These incubators may be used to transfer technologies to the country, and even if they are not invented in the national research and innovation system, they will have many economic benefits. Understanding a technology that is not present in the nation's industries is economically productive and is related to transferring high-tech technical skills and science to factories, production lines, or the service sector in the country. We cite two examples here for illustration purposes only. The first is a drug invented 100 years ago, which has no intellectual property rights; however, we still import it at a market cost of more than ten billion Saudi Rials. Establishing a company that transfers this knowledge to production lines inside the country to obtain part of the market share for this drug is economically beneficial at the national level. Another example is a chemical invented more than 200 years ago (with no intellectual property rights), which is used by the chemical laboratories inside the country; however, we still import this chemical at one billion Saudi Rials. Establishing a company to manufacture this material, even with a small factory, to meet national needs, or at least to acquire a share of the national market, is a good practice and has clear economic value.

These examples help us understand that the entrepreneurs and founders of (non-innovative) technical companies are often specialists in a specific field, such as medicine, nano-science, space, or nuclear science. The accumulated knowledge balance in the education of some of these individuals has not yet reached the stage of economic success. Technology incubators are an accelerating factor for converting the scientific knowledge of these individuals into investments with returns for the country and for the individuals themselves. This indicates the importance of establishing technology incubators, even if their cost is higher and the number of their beneficiaries is less (i.e., a small segment).

As for the hybrid incubators mentioned in the reports prepared by the Organization for European Economic Co-operation (OECP), these incubators incubate anything (with two or more types of incubation categories), which may not be from the same field. We should not support their presence in the country due to their lack of specialization. This is inconsistent with

the thesis presented by Adam Smith in the book The
Wealth of nations in the eighteenth century, in which
Adams notes that the basis of economic prosperity in
the Industrial Revolution was the division of labor and
an increase in specialization, with his famous example
being a needle factory.

Chapter 13

Venture capital: the nucleus for generating the nation's wealth

We do not know when the concept of venture capital emerged, but it is related to the sea industry and sea voyages, where the venture-level in each voyage was very high, with four common scenarios: the ship sank, lost its way, fell prey to sea pirates, or returned safely with flying flags. Even if the ship returned safely, the amount of real gain remained unknown, as the gain may have exceeded that recorded in the financial books. Venture capital investors receive their shares upon the liquidation of assets at the end of the voyage. Therefore, the sea industry was the first industry with venture capital in the past. However, in order to attain an unconventionally high profit, many enterprising merchants took to the sea or handed their trade to a crew, a ship owner, or marine caravans.

In ancient customs, the owner and crew of a ship owned a fixed share of the goods the ship carried. A reference indicates that a (quintile) custom 20% dedicated to the ship owner and crew was prevalent in European countries in the Middle Ages. It was common for some wealthy (people who do not trade) to finance a trip in agreement with the share allocated to them. Those financiers were adventurers who risked their money while knowing that loss was likely; however, they desired to accumulate their wealth and double the amount risked. This was their prime motive.

Something similar, albeit with less risk, was happening to caravans of land trade, despite the fact that the land is safer than the sea. Among the Arabs, many tribes used to protect land caravans as a profession and a source of livelihood for their followers in the geographical areas under the control of each tribe. The practice was done to allocate a portion of the caravan money (or profits) to the tribes of each geographical zone that protected it. Bandits were spread in the mountains, valleys, and deserts, threatening the safety caravans to reach their destinations.

Therefore, the concept of venture capital has existed since time immemorial, even if it did not have that name. Let us track how this concept developed in modern times. The modern beginnings of venture capital coincide with the beginning of the industrial revolution, along with the emergence of trains, railways, and heavy industries. In the sixth decade of the nineteenth century, a group of bankers and wealthy families in France, Britain, and America funded projects in the steel and railway industries with tangible assets that could be acquired and seized if the project failed, which was a way to collect a portion of the capital spent on the project. An example of this was in 1854, when three wealthy Americans funded the construction of a railroad that crossed the American continent from the Atlantic Ocean to the Pacific Ocean. Then, the idea of co-financing projects with large capital emerged, which bore the risk of being involved in very unfamiliar industries at that time. These investors bore the risk of setting up such mega industries and projects.

During World War II, the Americans and British faced a surprise when the Germans destroyed all aircraft in the first campaigns of the war in 1942; however, not a single plane hit German soil, and no target was planned to be bombed before the campaign. Upon investigating this tragedy, the Allies found that the Germans had a series of advanced fixed and mobile radars on their aircraft, which was uncommon at the time. This advanced German technology made the Americans reconsider their research capabilities and technical development in radar science. They then established a secret radar laboratory at Harvard University in 1943 with a capacity of 800 scientists, specialists, and technologists. After devising a solution involving covering their aircraft with aluminum materials, the American planes were able to disable the German radars and hit their targets on German soil.

After the end of the World War II, America worked to remain technically advanced, and the US Department of Defense allocated funds for universities to develop radar technologies and new weapons. Since universities are not factories, the university professors and students established those factories and companies after using solutions and innovations developed primarily at universities. This comprised the nucleus of the start-up technology companies established in the Silicon Valley. The first of these companies was funded with government capital from the Ministry of Defense. Professor Friedrich Terman, director of the Harvard's secret radar laboratory during the war and president of Stanford University during the postwar period, played a role in facilitating the creation of these first companies with university students and professors. Then money started to flow from some wealthy families in America to establish technical companies to serve agriculture, machinery, fertilizers, and pesticides in various other technical fields.

The first specialized venture capital company was established in 1946 with government capital, under the name of the American Research and Development Company (ARD), founded by Georges Doriot, Dean of the Harvard Business, who later founded the famous French business school, INSEAD. Then, in the 1950s and 1960s, the establishment of venture capital firms from investor money began, including the DGA in 1958, Rock and Davis in 1961, and Venrock in 1969.

Then, there was a jump in the liquidity injected into the venture capital sector when the United States Pension and Retirement Fund allocated 10% of its assets to venture capital in 1979. In one year, the amount of liquidity in this sector increased tens of times, which was only the beginning.

In 2018, the market in the venture capital sector (around the world) amounted to USD 254 billion compared to 174 billion in 2017, a rapid growth indeed. Some financing deals as venture capital for some companies even exceeded one billion dollars. Among them are the American cigarette alternative company JUUL, which received USD 12 billion and 800 million in 2018; the Chinese company Bytedance, with USD 3 billion in 2018; the Singaporean company Grab, with USD 2 and a half billion; and the Indonesian company Tokopedia, which received more than USD 1 billion in venture capital in 2018.

However, we find that the volume of the venture capital market in Saudi Arabia did not exceed USD 50 million in 2018, according to a report prepared by STV.

An insight into the venture capital market in Saudi Arabia indicates that despite the fact that the market is very small and does not constitute 1/1000 of the global market, there are signs of good beginnings. Many boards of directors of major companies operating in the Saudi capital market have started to establish small subsidiary companies devoted to investment in venture capital, including some banks, oil firms, and telecom companies. In addition, some entrepreneurs who received support from venture capital in their beginnings have also played their part in setting up venture capital firms.

Entrepreneurs, innovators, academics, and specialists in specific technologies have had promising small-scale ideas. Those who currently possess wealth or have the ability to direct these funds or capital from owners of large companies or state officials think of these ideas as simple, naive, or even childish. These ideas, however, might be able to generate tremendous wealth in the future. Video games, for example, have become a market in excess of USD 150 billion worldwide. Ideas for very simple applications that embody the desires and hopes of children or a group of consumers may thus become an engine of the economy, a source for increasing the country's wealth, and a model for other nations. Digging deeper into a study of the market, trade and investment will not lead the wealthy to pursue these ideas because many of them are simple in essence and connect to the concerns or needs of a segment of consumers or beneficiaries to which the investor may not belong. Therefore, in order to avoid missing an opportunity, it is necessary to harness funds to finance these startups, regardless of the market studies of these ideas, because some of them will create new markets that those studying current markets cannot absorb. Even forecasting may not work as an indicator to identify which startups are worthy of financing. Thus, we must expand our foundations, reduce restrictions, and allow venture and error. Some reports indicate that 2/3 startups disappear after a few years.

We return now to linking the subject of venture capital companies to the national research and innovation system. In order to allow startups and companies based on innovation and new or transferable technology, we need to establish venture capital funds in every academic entity and allocate annual funds from the budget of those entities to feed the funds with no more than 10% and no less than 3% of the entity's budget. These funds will finance startups via the faculty and students of the academic entity, with a proposed ratio of two thirds for the faculty and one third for students and fresh graduates (a maximum of three years after graduation). Organizationally, these funds should be linked to the head of the entity, such as the Chair of the funding board of directors, while appointing a dean or vice-chancellor for this position. There should also be regulations and controls to enable the faculty to establish startups and give them time without causing the loss of their jobs for a period not exceeding three years. As an example of this, academic bodies may add something similar to the "sabbatical leave" program given to faculty members in universities—one year for every five years of work. This might be called "Entrepreneurship leave". This will reduce the risk to entrepreneurs in the academic sector and enable them to succeed. An academic entity must reserve its rights by owning a small percentage that does not exceed 15% in a startup in exchange for financing. For students and fresh graduates, there may be a competition every semester or year, where the best business idea will receive an allocation for the establishment of a company based on the winning idea.

The allocation of a portion of retirement and social insurance funds to venture capital not exceeding 20% of investments of the retirement and social insurance funds is a good step forward. Everything mentioned above involves the injection of liquid venture capital from a government-funding source. Despite the importance of governmental funding, which has proven successful in some countries, such as China and America, encouraging investors to fund or start new venture capital firms is extremely important. The establishment of a venture capital chamber corresponding to the chambers of commerce may push the wheel forward by offering a package of benefits and facilities for venture capitalists, including tax exemptions and accelerating government management procedures, such as establishing a government channel or a path to expedite their actions to highlight their importance.

To conclude this chapter, without high liquidity provided as venture capital, tens of thousands of ideas for leading business companies, which may create a boom in the country's economy that exceeds the oil boom, will not see light.

.

Chapter 14

Technology Licensing Offices: Key for the Research Feed Industry

Technology licensing offices, under various names, from a research entity to another or a country to another, such as technology transfer offices or research commercial centers, such as those at Oxford University, are management units that manage the intellectual property of a research entity and contribute to the transfer of technologies and innovations from research entities to companies and factories for a royalty that they charge for granting the right to exploit their intellectual property.

Historically, Stanford University established a technology licensing office in 1970, followed by some other universities, until reaching its climax in America with the emergence of the Bayh–Dole Act in 1980, when legislation allowed the ownership of any intellectual property rights from federal funding to go to the university or research entity that obtained the intellectual property. After that, most universities in America started to include technology-licensing offices to dispose of intellectual property and manage it in order to collect royalties from intangible assets. This system abolished the previous system that had been in effect since 1950, which stated that intellectual property royalties arising from federal government funding are a right of the federal government, which the research entity has no right to dispose of.

Technology licensing offices work as a channel between academia and industry, and their tasks have two categories: internal, with innovators, and external, with investors. Internal tasks include attracting researchers to present novel or strange ideas that may be innovative and then evaluating these potential ideas and innovations and completing the legal procedures for preserving the intellectual property of these potential innovations. The external category of their tasks includes marketing the innovations of the RE protected by intellectual property rights, communicating with investors and companies to show the value of these innovations, preparing contracts and agreements for licensing exclusive and non-exclusive rights, and negotiating the distribution of royalties from the profits of these innovations. Some of these tasks may interfere with the management of innovation financing and others in the creation of startups emerging from these innovations.

One role of these offices is to educate researchers about the importance of preserving intellectual property rights for their potentially unprecedented work. A researcher is not an investor; his main concern is scientific publishing, and the process of preserving intellectual property rights hinders or delays scientific publishing and may not yield revenue. Therefore, these offices suffer greatly in the process of educating researchers and identifying the benefits a researcher may gain from disclosing a possible innovation for a research entity. These attempts include persuading the researcher to turn his knowledge into something tangibly beneficial to society, which does not include published and specialized scientific research in a specific field. These are works written by specialists in a specific field for professionals in the same field to read, like a code that only specialists can decode. As part of these attempts to persuade researchers, these offices try to help researchers see the outcome of their scientific work in the form of products, goods, and services sold in the market. The promise of gaining revenue, however, is not always feasible or convincing for researchers.

These management bodies that exist in research and innovation entities have many benefits that help bring wealth to the research entity and may become a constant source of income for a research entity. When properly operated, they may transfer many technologies and innovations from the university or research entity to the market through the company that will license this technology. Upon the success of these management bodies, researchers and faculty members in universities will not need to leave their jobs and devote themselves to trade, investment, and entrepreneurship. Their ideas and innovations will see the light, move to industry and bring them profits while they remain in place.

To demonstrate some of the successes that technology-licensing offices can achieve, we refer to Stanford University's Technology Licensing Office for its close relationship with Silicon Valley companies. On its website, the office provides investors and companies with more than 1,500 technologies produced by the university and protected by patents. Companies can see and select the appropriate inventions when they want to import technology from the university. In 2018, the office generated nearly USD 41 million in revenue from 813 technologies. It is worth noting that the revenue from one of these technologies exceeded USD 11 million in one year for the university. There are many successful models in the area, including the Technology Licensing Office at the Massachusetts Institute of Technology, which provides a portfolio of software produced by the university and available for licensing to companies and investors. It is worth noting that this office has about fifty employees comprising financial and legal specialists, marketers, and information technology operators, in addition to the main staff who are in charge of attracting innovators from the university and marketing innovations to the companies.

Chapter 15

Enabling Innovative Researchers to Be Entrepreneurs

In his book, Democracy, Socialism, and Capitalism (1942), Joseph Schumpeter, Austrian Minister of Finance and Professor at Harvard University, provided what was later known as the "Mark II" model. He argued that the fundamental innovations systematically pumped into the national innovation system come from large organized companies and major entities that contain frameworks, models, processes, and procedures that make innovation easier because of substantial financial support and a professionally advanced work team, allowing these entities not only to quickly innovate but also to convert these innovations to products sold in the market in relatively short times. However, we are inclined to accept the Schumpeter model "Mark I", which Schumpeter put forward in 1912, which suggests that the economy is built mainly by the efforts of startups and creative individuals who have struggled to invent new products and services and to convert them into goods sold in the market.

Schumpeter's "Mark II" model produces sustained progressive innovations for the continuation of existing sectors, while the innovations of individual entrepreneurs in the "Mark I" model revolutionize markets, promote new technical generation, and destroy the old commodity markets, as Schumpeter himself pointed out in the theory of destructive innovation. This is consistent with what Clayton Christensen proposed in his book, The Innovator's Dilemma, in 1997.

The bottom line is that radical innovations that create booms in the market arise from individuals and are not directed at the request of large entities. Often the systems and procedures that the best research and innovation entities possess around the world cannot be recognized or monitored in their early stages. Therefore, an innovative individual who engages in radical innovation that will create a new market alongside market studies that technology licensing offices in universities conduct, for example, cannot measure the market value of a commodity in a market not yet created. Herein lies the difficulty for an entity in identifying these innovations and for innovative individuals who do not receive support at the start of their innovation.

These innovators always have only one option, which is taking risks as individuals, sacrificing their limited financial resources, and borrowing to finance their innovative projects driven by a profound belief in the success of their ideas. Despite all that, they may not succeed because their resources were not sufficient to allow their innovations to reach the market or because their experience and practice in investment and trade is very limited, requiring that they take a few years to understand and overcome the first obstacles.

In all cases, even if we do not find a way to discover and identify these technologies in their early stages, we need at least to develop means that enable these innovators and those willing to take risks to obtain support in some way. Although these innovators may be from outside the research and innovation system itself, we still believe in what is produced by laboratories and experiments. Many scientists and researchers try to highlight aspects of their work that they have previously seen on a small scale in their research, but they often do not succeed in highlighting such aspects in the early stages because of their weak marketing capabilities or due to their profound knowledge in their field compared to common knowledge.

Here, we will mention some examples to show innovations from individuals that have created new markets, some of which came from individuals belonging to the research and innovation system in their country.

The first story: Bushnell and Daphne established the Atari video game company in 1972, which was a revolution that created a new market for a new commodity. In 1977, they introduced the Atari VCS console (later rebranded as 2600), which that sold more than 30 million units, and the company earned more than two billion dollars. Many subsequent companies with sophisticated organizational structures and huge budgets in the video game industry entered this market until sales in the global video game market reached USD 152 billion in 2019. There is a competition between various countries, such as China, Japan, and America, for the lead in this market. This market would not have been possible without the industry's first innovative entrepreneurs. At that time, the risk was too high to create a commodity that had no market and for which there was no need among potential consumers.

The second story: Beckman, a professor at the California Institute of Technology, invented a pH meter in 1934. He established a company to manufacture and market this product that had no counterpart and no market. Today, almost no chemical or physical laboratory around the world does not have a pH meter. Also, in the 1950s, Beckman invented a spectrophotometer for invisible UV rays. Although there was no available market information for this product, one of the Nobel Prize winners in chemistry, Bruce Merrifield, said in 1984 that it was the most important scientific invention of its time.

The third story: Philip Sharp, a professor at the Massachusetts Institute of Technology and a 1993 Nobel Prize winner in medicine, discovered a method for RNA splicing and established a biotechnology company in Geneva in 1978 (biogen) with his colleagues to market this invention. The assets of this company are currently estimated at more than USD 24 billion.

The fourth Story: Mark Horowitz, a professor at the College of Engineering at the Massachusetts Institute of Technology, developed a memory-capacity technology and established, with his colleagues in Rambus Inc., Rambus Ram in 1990. The profits of this company currently exceed USD 300 million dollars annually. This company has developed a memory-optimization technology for many applications and has licensed it to Intel, Microsoft, and IBM.

The fifth story: Patrick Hanrahan, a professor at Stanford University, developed interactive visual data in 2003 in partnership with his colleagues at Tableau Software Company. The profits of this company exceeded USD one billion in 2018.

Chapter 16

Investors' Patriotism May Stimulate Investment In the National Research And Innovation System

What is the relationship between an investor's patriotism and the national research and innovation system? Profit is profit, whether it is an outcome of national or foreign research and innovation systems. Why, then, should an investor prefer the national research and innovation system even though he can obtain something similar, if not better, from foreign research and innovation systems, especially in the absence of benefits provided by the state? Investors certainly know that the level of maturity and readiness of foreign research and innovation organizations is higher than national offerings.

Pumping financial resources into the national research and innovation system leads to the prosperity of this sector and its expansion by creating new jobs, laboratories, experimental fields, and research facilities, which, in turn, will lead to a greater demand for goods and services to meet the new needs of this expansion. This will produce an internal market, and increasing jobs in this sector will lead to the increased consumption of job recipients to direct this consumption locally. This is an integrated economic cycle that will eventually lead to an activate economy and an increase in gross national income.

If this investor lives in the country, then his sons, relatives, tribes, and the people of his village or home city will benefit from this process of pumping money into research and development at home rather than transferring it abroad. This investor, even if he is ultimately interested in profits, may have national pride for the benefit of his country and his family, which will ultimately be beneficial.

If an investor abandons this practice and insists on transferring funds to foreign research and development entities that outperform their local counterparts, he is unconsciously feeding and revitalizing the economies of other countries and contributing to finding jobs and markets for goods and services that serve research and innovation systems abroad. Is it not better to pump this money into one's own country?

In Saudi Arabia, almost 80% of what is spent on the national research and innovation system comes from government funding, and the private sector contributes nearly 20%.

Of the 20% spent by the private sector on research and innovation, most of it is harnessed for foreign research and innovation systems. Even what is spent on research and innovation from the private sector is not spent within the country.

In South Korea and Israel, three quarters of expenditures on research and innovation come from the private sector, and most of these funds pour into their own national research and innovation systems. Both spend 4% of their national income on research and innovation, compared to 0.8% in Saudi Arabia.

The bottom line is that investors should direct the private sector to invest in national research and innovation entities even if the state does not offer them a package of benefits and facilities because the payoff is ultimately worth it. In the long run, they will earn higher profits, even if not in the short term due to the lack of maturity of many national research and innovation entities.

PART THREE | Innovation: A Pillar of National Security and Economy

Ingredients of National Economic Growth: Labor, Capital, and The Missing Element

Adam Smith, in the Wealth of Nations, and Karl Marx, in Das Kapital, have made it clear that the economic growth of a nation depends mainly on two elements, capital and labor, which are the main inputs into the economic ecosystem of a nation.

Then came Robert Solo, who explained in his many articles the importance of the missing element from the previous equation: technology and innovation, which he called "technical progress." Professor Solo of Massachusetts Institute of Technology received the Nobel Prize in 1987 in economic sciences. He made it clear that the economic outputs of an economic system through innovation (the new capital) are more productive and of higher value than other types of output.

In order to link things together, we return to what Thomas Friedman presented in his book, The World is Flat and the theory of the third globalization. Through changes in the complex global supply chain over the past few decades, the United States of America has developed a strategy to keep generating IP products and shifting services and manufacturing to other countries, such as China, India, and Mexico. Some western European countries have followed suit. Many American products are not produced inside America, even though they are American innovations and designs. For example, Chevrolet and Chrysler cars are made in Mexico. Apple iPhones are manufactured in China. Ray-Ban glasses are designed at the headquarters of the American company and manufactured in Italy and China. Many American tire factories manufacture their products in Russia and Mexico.

Some believe that the transfer of these industries from the United States of America to other countries weakens the USA economically and marks the beginning of the end of the country's economic strength. Friedman suggests that this is a strategy that was engineered to transform the low-level supply chain to require only production (manufacturing activity without thinking) or the implementation of services for other countries, while the state maintains two things: the intellectual production of new technologies and innovations, which in turn guarantee most of the net profit, while the rest of the profit goes to cheap labor wages in countries abroad, which bear the consequences of environmental pollution. Added to this is the capital market, as all of these commodities are sold and purchased in US dollars, which also guarantees the survival of American power.

In this way, the importance of caring for the national research and innovation system becomes clear, as this system will guarantee us the most important element in the equation of our economic growth. The shift to intensive industrialization or the provision of distinguished services will not place us at the forefront of development without "technical progress", through which we will contribute in making industries and services of higher economic value and better productivity. We do not mean here by a direction of "technical progress" that we should import technologies and innovations from abroad but rather that we should produce them at home through the national research and innovation system.

Chapter 18

Innovation Is the Source of Homeland Security and Defense

About five thousand years ago, the ancient Sumerians in Iraq discovered that adding tin in a small amount to copper makes it one hundred times harder than iron; this substance was bronze. The discovery of bronze led to the transformation of the war industry for the Sumerians, who made many of their weapons, including swords, daggers, shields, helmets, and spears, out of bronze. Based on this technology, the Sumerians were able to invade the neighboring peoples until they became an empire. All this success was due to their at-home advancement of metal sciences and how to convert this knowledge into successful practical applications, leading to great military power. This knowledge also led to the development of a higher standard of living and well-being for the Sumerians, who used this new material to manufacture cooking pots and utensils and other everyday tools due to bronze's durability and longer sustainability compared to their traditional stone and copper products. When more vulnerable nations began to learn the technical secrets of bronze, the Sumerian civilization began to decline because younger generations could not find new technologies and innovations to distinguish them and give them a competitive advantage over other peoples.

History provides us with a great deal of evidence that technical and cognitive progress and innovation are the primary makers of civilization. When the Portuguese learned their navigational skills and began to use Arabic tools related to the sea, which the Arabs developed in Andalusia, including triangular sails and astrolabes, they managed to explore islands in the ocean. After discovering the rifle "musket" produced by the Ottomans in 1465, the Portuguese used such rifles produced in Malaga (Othman island) to conquer the aboriginal peoples on the newly discovered islands and overseas, followed by the Spanish, the Dutch, and then the English and French.

In 1543, a ruler of a Japanese island bought two rifle muskets from Portuguese merchants and ordered reproductions of a similar weapon. After a series of adjustments in the speed of loading and ease of carrying the rifle, the Japanese produced their own model, "Tanegashima", named after the ruler who ordered the manufacture of this weapon. In 1575, a unit equipped with 3,000 Japanese rifles managed to defeat their enemies at the Battle of Nagashino. Then came the most important event of the Japanese–Korean War in 1592, when the Japanese managed to achieve a decisive victory by defeating the Koreans in just 18 days and occupying Korea; this was all due to 160,000 Japanese soldiers possessing rifles. Modern technology thus enabled the Japanese to prevail. The Japanese remained a military power in East Asia until the Second World War. Professor Delmer Brown of The University of California at Berkeley published a scientific paper in 1948 entitled "The Impact of Firearms on the Japanese Wars", in which he concluded that the Japanese ruler did not buy the two rifles at a very high price to use them or to acquire what is modern or beautiful but in order to reproduce an unknown weapon for home use.

With the start of World War II, German technical progress was clearly significant in all industries, both military and civilian, with its advanced means of supply and logistics. Due to its technical progress, Germany managed to overtake almost all of Europe without much resistance in less than three years. Germany's advancement in radar science is an example of this technical progress, through which they managed to crush the British Royal Air Force, as well as American fighters and bombers.

The United States of America did not overcome Germany's advanced technologies until it focused later in 1943 on developing technologies and innovations against these advanced technologies. America prevailed over Japan only due to its progress in nuclear science, through which it defeated the Japanese with its innovative nuclear bomb. As a result of only two bombs, Japan surrendered in 1945, putting an end to World War II. These two bombs caused the deaths of more than 100,000 Japanese citizens in Hiroshima and Nagasaki.

During the Cold War, the technical and innovation conflict between The United States and Soviet Union reached space, which became a battleground for the conflicting nations. Those who innovated and provided better technical solutions were able to maintain their progress and dominance. Whenever US intelligence discovered that the Soviet Union had developed a new technology, that technology was sent to research centers to develop a counter-weapon; funds were pumped from the state treasury to feed this research. These researchers developed prototypes and tested them through laboratories and other experiments. The Soviets did the same when they discovered new American technologies and innovations. Both parties turned to their research and innovation systems to excel during this conflict, and investments in research and development were high on both sides. There was significant progress in the underwater armament systems, submarines, the nuclear field, and jet engines for fighter fleets and spy satellites. It is worth noting that the Soviets produced approximately 100 million pieces of an automatic rifle weapon known as the AK 47 "Kalashnikov", which was invented by the Soviets in 1947 as the first continuous-fire guns.

The struggle between nations is decided by technical progress and innovation. Innovation is not merely a requirement for economic growth, social welfare, and improved livelihoods. Innovation is an element of national security and a safeguard for the peace of nations.

Even if some countries precede us in a scientific advancement or field of knowledge, this does not mean accepting the fait accompli and turning a blind eye to the bright future that awaits us. Our focus on looking after and caring for the national research and innovation system is an imperative requirement to protect this nation. The advancement of the Germans in radar science did not prevent the Americans from developing techniques to disable radars, and the production of nuclear bombs by the Americans did not prevent the Soviets from producing the nuclear bomb. In this way, scientific progress must continue. We should not take for granted that only some countries have the upper hand in technology or weapons because we will be able to produce better or equal technology in the future.

Chapter 19

National Security Research and Innovation Council

There is an urgent need to direct national research and innovation entities with a civilian character to develop weapons and army equipment in times of peace so that we can defend the homeland in times of war. Efforts must be directed and united to promote innovation in defense technologies in various fields, from satellites and radars to submarines, ships, aircraft, missiles, and armored vehicles. For this to happen, it is assumed that there will be an umbrella to help guide and oversee these efforts without interfering in how research entities operate but instead providing them with the necessary liquidity, specifying what needs to be developed or innovated, and guaranteeing the transfer of the best outcomes from the national research and innovation system to defense factories. Funding research and innovations in security and defense are, however, outside the aspirations and endeavors of the private sector.

This council should report to the highest authority in the country to give power and influence to link various sectors to achieve their desired goals. Although some basic defense factories currently exist, the national research and innovation system has made no contribution to the innovation and transfer of its technology from laboratories and experimental fields to factories. These technologies in factories were transferred from outside the country, and most of them do not stipulate what is made in countries that have exported their technology to us. It is common knowledge that foreign entities export technology that they have already surpassed.

Although we have purchased some of the best types of weapons from various countries, we are not supposed to continue importing such weapons and are also not permitted to produce them. Even discounting prohibited types of arms, such as nuclear weapons, what about the rest of these weapons? What if our financial resources are depleted? What if an economic boycott occurs between us and some of the countries that sell us our best weapons?

Weapons are products that evolve and improve over time. What we buy or acquire today will not have advantages and characteristics beyond what is produced tomorrow.

Science is advancing and weapons are developing. So, is our strategy always to buy? For manufacturing, the weapons given to us to produce our own weapons have been technically surpassed by those who allowed us to produce them. In other words, we need to rely on our own capabilities in innovation and technical development to develop what we need in the future.

The council we referred to above should be a permanent council in peace and war. In times of war, we propose the creation of a temporary council (that will continue until the war ends), which will undertake the process of research and innovation for solutions faced by the army on the front line. This council will determine every danger or difficulty in the military field by communicating with the leaders of the Ministry of Defense and then mobilizing teams of scientists and providing them with funding to develop solutions to overcome these difficulties. Then, this council will hand over these solutions to the army or establish production lines for these innovations and technologies that embody field solutions to problems and challenges in the field. In his book entitled Science: The Endless Frontier (1944), which was dedicated to the former President of the United States of America, Franklin Roosevelt, Vannevar Bush explains how the Scientific Research and Development Office that the United States established during the war (which Bush managed) worked organizationally by being linked to the Head of State. Vannevar reminded Roosevelt that his victory over the Germans was due to the scientific developments that scientists contributed to the field of radar, while the victory over the Japanese was due to scientific progress in the nuclear field.

We need to develop a similar model or better. The need for self-sufficiency in arms is necessary to defend the homeland. Most important is the development of solutions and technologies for the problems we face directly during a war. Weapons developed before a war may not meet the needs at time of war—not in terms of their numbers but in terms of their quality—since an adversary may develop novel weapons and surprise us after the start of the war.

Chapter 20

National Competitive Advantage: Based on Innovation

About forty years ago, Professor Michael E. Porter of Harvard University created a concept that summarizes a firm's ability to compete with its peers in the same business and geographic region. From this research emerged the concept of "competitive advantage", which justifies the ability of a company to outperform its competitors in the market. Using an analytical study of the competing companies in different markets, Porter explained this phenomenon as the added value of a company over that of other competing companies. He concluded that price competition is harmful to the economy and trade without adding value to the goods and services provided by the competing companies. The impact of this competition is initially negative for competing companies, even if clients benefit from the price difference temporarily. However, in the long term, clients are affected by the lack of progress and development of goods and services provided to them.

Healthy competition between companies occurs when each competitor adds value to its product, which causes a company to gain a competitive advantage against its peers. In summary, how can a company add value to the goods or services it provides to clients? This can only be done by developing, innovating, and improving the product, commodity, or service, thus providing added value to the consumer and the economy as a whole, especially if every competitor is trying to add value to its goods or services. With the passage of time, products and services in the market will become more advanced, developed, and beneficial.

In "The Competitive Advantage of Nations" (1989), Porter expanded this concept to include defining the competitive advantage of countries by applying the same previous concept to competing companies in a specific market. He studied the value chain of ten countries to learn about their competitive advantages. He concluded that states, like companies, can develop, improve, and innovate in order to excel and prevail in the fields of defense and war or civilization.

It is this ability to innovate, develop, and continuously improve that will give our country a competitive advantage over other nations. Innovation, in general, is a product of national research and innovation system outputs. This brings us full circle: what will give our country a competitive advantage is a strong national research and innovation system. It is true that there are opportunities for innovation that do not result from this ecosystem, but these are disorganized innovations and may be interrupted because they are the efforts of individuals and not institutional work that ensures sustainability and continued innovation and development.

This does not mean that focusing on the national research and innovation system will only give us a competitive advantage for research entities in distinguished research publishing. This is a catastrophic mistake. Science is not for its own sake but for its benefits. Even if the progress of understanding is relegated to a narrow field, it is not for the sake of science but for the sake of that field that we increase our awareness of it. However, the national research and innovation system is meant to serve all sectors, not only itself. Excellent petrochemical research, for example, will inevitably develop the petrochemical sector and all companies that operate in this field or intersect with it, leading to their prosperity and the introduction of competitive advantages for their products and services. In this way, the advancement of science will create competitive advantages for the nation in all non-research activities of the state.

Chapter 21

The Nation's Value Chain

When we eat delicious food, we taste it, but we do not know how the chef came to produce such delicious food. We only acknowledge its taste and do not know how to produce it. When we track and monitor the chef, the secret to this food may not be the skill of the chef but rather the chef's cooking inputs (for example, using meat from lambs raised in good pastures). Competitive advantage here refers to the delicious taste of the meat, but the critical activity that led to this competitive advantage is using a type of lamb raised in good pastures.

Michael Porter singled out a chapter in one of his books to describe the concept of "value chain". In this chapter, Porter tracks how a company produced added value and gained a competitive advantage by analyzing its activities, inputs, and operations to determine the activity that led to the production of the added value on the business as a whole.

The specific critical activity (or activities) that are key to creating added value for an entity should be investigated, studied and identified. Then, other competing entities can replicate these activities by introducing their workflow steps and procedures to obtain the same competitive advantage that the strong competitor has and thus turn economic profits in their favor. However, this is not the ultimate goal in recognizing the strength of a strong competitor, but rather developing or improving the specific activities or creating an activity or process that is superior.

Since countries have competitive advantages, the value chain can be traced to identify the activities and practices that have led to obtaining those advantages. Hence, teams must be established at the national level to determine the competitive advantages of other countries, how those countries attained them, and to study the activities and procedures that led to said advantages. After that, the relevant activities and procedures should not be imitated but rather developed and improved upon. In this way, over time, we will obtain many types of competitive advantage that are difficult for others to keep up with. If a competitive advantage that we have become known to others, then the solution is always to find alternatives that surpass it. The march continues without stopping. When we are satisfied with what we have, this process will end.

To increase the practicality of the solution, tracing the value chain to the competitive advantages of a country is not as easy as tracing the same for a company or a non-profit entity. The ramifications of this process are many, and the related activities are numerous and intertwined in a complex model. Even more difficult is the lack of full information about the relevant activities and the difficulty in following their sequence. Therefore, this task is not simple and needs capabilities and talents to be allocated and focused to obtain a value chain for each single competitive advantage of a country.

Although the terms competitive advantage, added value, and value chain are modern and date to a maximum of forty years ago, the concept itself has existed since ancient times under different names. Just as makers in their industries were competing and trying to distinguish themselves from the others, so to were countries more eager and competitive.

Historical examples include the attempts by Muhammad Ali Pasha, the ruler of Egypt in the nineteenth century, to send ambassadors and missions to stay for years in the countries that he believed to have something unique to offer Egypt. Muhammad Ali Pasha organized embassies and missions to determine how the country advanced and continued to advance in a specific field and to attract the related sciences and skills. One of the books that explains these attempts was written by Rifaa Al-Tahtawi, who was a supervisor of the Egyptian mission to France. In his book Paris Profile, Al-Tahtawi wrote down everything that he believed to be distinctive in Paris and wanted to transport to Egypt, explaining how each element was unique and what its transfer mechanisms were.

This is one example of rulers' keenness in modern and ancient times to make their homelands the most prosperous and urbanized, as well as the most secure and able to defend themselves or attack. Many confidential historical attempts sought to identify the competitive advantages of other competing countries, determine their opponent's strengths, and find ways to obtain similar strengths. Not all such historical events, however, were written, transmitted, or disclosed.

Chapter 22

Basic and Applied Sciences: Directing Country Resources

S cience involves the creation of new knowledge by exploring the unknown. For us, the unknown may be natural phenomena and an explanation of the movement and presence of things around us (such as the basic sciences). Other unknowns may relate to solutions to specific problems we face and applications for what we know about the phenomena of nature to improve our ways of life or overcome difficulties facing us (such as applied and industrial sciences).

All applied sciences originate from the basic sciences and do not exist independently in isolation. The progress of any applied science is always based on equations, laws, or rules derived from a basic science. However, the general public often sees that useful items come from applied science, a belief that often places pressure on governments not to directly financial resources or care for the sciences that the public sees little use for. Although all applied science branches rely on and derive their knowledge from basic science, the general opinion formed in all countries is to dry out basic scientific resources or limit them as much as possible. On the other hand, we find that private sector investments around the world go to applied sciences, but only those closest to industry are funded. This leaves the basic sciences without funding. This is an ongoing challenge facing those in charge of formulating policies and directing resources in the national research and innovation system.

About 2,300 years ago, Aristotle realized this fact when he found that his excellence in medicine offered him many financial resources, which led him to establish a research academy. Then, he redirected the resources he gained from his knowledge and mastery of medical science to other basic sciences.

Humans often aspire to obtain quick and immediate profit from their investments and harnessing of resources, while few realize the inherent importance of basic science.

In his letter to President Franklin D. Roosevelt in 1944, Vannevar Bush noted that the military sciences and applications used and developed in the war were due to two factors: The first is the accumulated knowledge of the basic sciences over the centuries (before the formation of the United States of America). The second is the knowledge that the United States of America accumulated in the basic sciences in the fifty years before the World War II. When the war came and scientists were asked to find quick and immediate solutions to the challenges facing the army on the front-lines, these scholars had to research the basic sciences first to devise solutions to the practical difficulties faced by the army.

Thus, we must not lose sight of the importance of the basic sciences and allocate the necessary resources to them from government resources in the national research and innovation system, even if they appear less attractive to the masses and of little interest to the private sector. We suggest that the basic sciences should receive at least half of the funding allocated to the applied sciences in the worst-case scenario.

PART FOUR | Urgent Solutions To Reform The National Research and Innovation System

Chapter 23

The Management Challenges for Research and Innovation Entities

I t is obvious that everyone who assumes an academic management position, be it a department head, a dean, a rector of university, or a director of a research center/institute, is a researcher in the first place. There are advantages and disadvantages to this practice. One advantage is that a researcher knows what his fellow researchers are going through, which makes it easier for him to understand the affairs of researchers and to communicate with them to ensure the system progresses smoothly. All researchers have a shared language that serves as a communication platform. This reduces the problem of trust between researchers and management because the manager is a researcher himself. When appointing a person who is not a researcher to such a position, all parties will suffer due the difficulty in understanding the nature of research work compared to work in factories and service companies, including the nature of the production of a research product, how a research idea turns into a proposal and a proposal turns into a study, and how this study is carried out before publishing its results. This process is not easy to understand for those who do not work in a research sector. We mention, for example, knowledge of the time required to produce the various stages of research, including writing a research proposal, writing a technical research report, and publishing a scientific paper. These stages have no fixed timeframe that an outstanding manager from outside the research sector can comprehend; such a manager may think that there are specific timeframes that researchers can be held accountable to by tracking and controlling them. It may take up to two years for scientific research to be published in a reputable journal after the completion of a research project and an analysis of its results. How does this manager (an outsider manager) remain patient during procedures he or does not know about? The only way to solve this problem is to reeducate the manager as a researcher, which may take years. Hence, the common practice is acceptable, which involves selecting researchers to assume management positions in academic entities.

What, then, is the problem with a researcher taking a management position in a research entity? There are two main problems: a lack of the necessary qualifications and management maturity for researchers to practice management, as well as competition for resources with follow researchers and a bias for his own discipline.

Through the experience and long-term observations of many who have held management positions in several research and innovation entities, many researchers have no awareness of the fundamentals of very common management concepts. They manage by intuition. We do not deny that some research managers are excellent, even if they manage by intuition, but the management principles and concepts that constitute a shared language for managers are not available to them. In the absence of a shared management language in the research entity, each manager has his own management concepts, which cannot be communicated well with his supervisors and followers because of the lack of this shared language. Added to this is missing the opportunity to adopt the best common management practices that can be introduced to the system. Even if the entity leader (the first official in the entity) adopts a new management concept or principle and wants to apply it to the entity as a whole, it will collide with many obstacles and difficulties.

To address this challenge at a high level for those who are in charge of the governance of the national research and innovation system, we must work on two things.

The first is to prepare management qualification programs suited to research and innovation entities and to establish a management and executive training career pathway for researchers who have the potential to assume these positions in the future. This would be a process for creating leaders in the research and innovation system without leaving anything to chance. Creating new leaders and executives in these entities must be a systematic process and involve institutional procedures. The quality of research and innovative entities depends on the eligibility of these elite. We may need to design this executive career path for thirty or forty years for junior academic employees and academic researchers. These future leaders also need refinement and improvement over time in tandem with emerging and developing issues in the science of management and leadership. We also need to establish an academy for leaders in national research and innovation entities that will become a counterpart to the Staff Command College, in which officers receive their education in military planning and strategy at the Diplomatic Institute that graduates diplomatic corps staff. This academy will design, implement, and develop an executive path for research and innovation leaders. Within ten years, only graduates of this academy would be able to assume leadership positions, from the head of a department to a president or rector of a research entity. This academy would have the records of all researchers, their development, and their performance. This academy would contribute to nominating researchers to higher positions in the management sector, even in research entities other than their own. We should design a package of benefits and compensation suitable for each management position, as the burden increases. In this case, we could benefit from the experiences of executives to serve and reform several research entities without requiring them only to serve the entities to which they belong to.

The second challenge involves the problems in appointing researchers as managers in their own organizations. This has caused us great embarrassments in the past. We must address this problem to reform research and innovation entities. This is more important than the first challenge. If we cannot solve it, we would prefer to appoint people who are not researchers to manage research, which is safer, despite its limitations.

Currently, the heads of academic departments, deans of colleges and their deputies, university rectors and their deputies, as well as the heads of centers and research institutes do two things: They compete over the resources available for their own research, which deprives their colleagues who do not have management positions of such resources. Second, they believe that their academic discipline and associated disciplines are the most important. Therefore, they often devote resources to what they believe is of importance.

This is a management catastrophe that must be addressed when reconsidering the governance of the national research and innovation system. We will now put forth some possible solutions to control this phenomenon. Research entities must stop all those who hold a management position in a research entity (even at the lowest level of management, such as the head of a department) from conducting research. A manger should devote all of his efforts to management work, organization, coordination, and resource management, whether for laboratories, facilities, research experiment fields, or grants. He must choose to either work as a researcher or as a manager of a research entity to prevent any conflicts of interest. Many complaints often come from fellow researchers who compete with the financial and non-financial resources the manager works with. A manager must stop his research work until the end of the management term or assignment, at which point the manager can return to the ranks of researchers. In order to control this process, it will be necessary to exempt everyone who practices research from management positions. A manager should be terminated if he publishes a scientific publication or conducts any research practice during his term, except for submitting scientific abstracts at scientific conferences, as these presentations only show previous experiences the manager may have accumulated.

We can control a manager's bias for his research discipline in two ways. First, a manager can sign a written commitment not to allocate any of the resources or benefits of the entity to his specialization and not to direct the system in favor of this research discipline or the research disciplines that intersect with or branch from it. This problem can be clearly observed: The research disciplines to which the heads of entities belong flourish by the presence of such managers at the top of the ecosystem because of their allocations of higher resources and benefits. This generates a sense of injustice among other researchers The second solution involves the establishment of a general national committee, which may branch off from the National Scientific Council (which we have mentioned above) to consider directing resources based on the competences of deans, university rectors, and the directors of centers and research institutes. This committee will have the authority to hold managers accountable and freeze their powers if any infringement is observed. This would be done with the aim of balancing the scales of justice for the advancement and growth of the national research and innovation system. Maintaining the system away from the whims of its leaders is a national duty. Through previous discussions with hundreds of researchers, we found that each of them believes that their research discipline is the most important for the country. After analyzing their opinions, we found that ignorance of other research disciplines and good knowledge of one's own research discipline gave birth to this opinion. If a researcher becomes responsible in a research and development entity, he will, in good faith, direct resources toward what the researcher believes to be important, which may not be more important than other disciplines. We must be careful in moving such entities away from vain desires, even if in good faith.

Chapter 24

Incentive Based Performance for Researchers

Currently, the various national research and innovation entities (governmental) are generalized, without specifying a single methodology or mechanism that distinguishes a single entity. In general, the main incentives for the employees of research and innovation entities are academic promotions for academic faculty. To clarify this matter for those outside this sector, academic promotions, for those hired as professors, include only two levels after appointment as an "assistant professor", which takes approximately eight to fifteen years for most faculties. Some do not wish to obtain such promotions because of their low financial benefits; another category of researchers provides consultations to the public or private sector and is not involved full-time in scientific research, and a third category includes those involved in non-academic management. Added to this are those involved in managing research and innovation entities and those who have the desire to devote themselves to management.

All research support team, including laboratory technicians, specialists, research assistants, junior researchers, and management support teams, do not have special incentives. They receive the benefits and promotions that apply to other state employees. Their performance is not related to the support provided to their research or other researchers.

A few other benefits were introduced in 2011 to improve the positions of researchers; some of these benefits were subject to interpretation according to the understanding of each university or academic entity, after they had determined that the related expenditures would be high.

We now put forward a proposal to link incentives with performance. To do this, we must first divide the workers involved in scientific research into segments and provide incentives for each segment separately. The main segments could be as follows:

1. researchers who publish scientific research related to solving existing problems defined by the entity;

2. researchers who publish scientific research not related to solving the problems of the entity;

3. researchers who develop technologies and innovations related to problems defined by the entity;

4. researchers who develop technologies and innovations not related to the problems and needs of the entity;

5. research assistants expected to be future researchers and research assistants not related to research. Each category could be divided into several sub-categories as needed.

For example, let us choose a category (researchers developing and innovating solutions for problems defined by the entity). Each solution developed by a researcher (or a research team) may receive one point for each innovation disclosure form, three points for each patent filing process, and ten points for each granted patent. Patent licensing should not be part of the point calculations for researcher. A researcher may receive a share of the expected royalties at a fixed rate according to his participation with the rest of the researchers and the amount of funding provided by the entity to develop this solution. The fixed annual bonus system should be replaced with the points calculation system obtained at the end of each year, with a monthly salary increase granted to the researcher in the coming year. Every ten points can be replaced by 1% of the researcher's basic salary without a maximum ceiling, even if it reaches

50% of the basic salary. These added bonuses should be consistent and ongoing. As for the supportive team that reviews the innovation disclosure forms for inventing, filing, and follow-up until obtaining the patent (as well as the marketing process for the patent to be licensed and manufactured), all this is not the role of the researcher, even if he is the innovator. This is institutional work that must be performed by specialized individuals with their own financial incentives and benefits. For example, each form should be reviewed by two people separately; if they agree that the form is not valid, the patent filing process should not be completed. Likewise, the annual bonus system should be replaced by a point system for each task performed by the supporting research team member. For example, the employee in the technology licensing office could be given, for each review of an innovation disclosure form, a point, and for each patent filling, five points, with 20 points given for a successfully obtained patent. Points are awarded at the end of each year, and the team member is awarded 1% of his basic salary for every 1,000 points, and so on. To market the innovation, the employee receives a percentage for each transaction signed with even 1% of the royalties. This could also be applied to employees in the research supporting team, including technicians and specialists, but the most important process is to determine the nature type of the researcher's work and the amount of time expected to complete each task and then translate it into points and calculate the amount of achievements during the year to determine his annual bonus. Consider a technical employee working with an inductively coupled plasma (ICP) instrument to analyze heavy elements. A point is granted to the employee for each sample analyzed; then, the points received during the year are calculated in order to award his bonus, and so on. This process requires multiple teams to establish the incentive program for each category, which cannot be detailed in a book or article. Accordingly, it is necessary to establish a committee from which several teams can branch out to study and revise the incentives and point system according to the principles referred to above.

Chapter 25

Incentives for Investors to Inject Money Into the National Research and Innovation System

The first thing an investor thinks about when he wants to develop or invent a technology is to turn to those who own the technology—i.e., the foreigners who imported it into the country. Accordingly, it is expected that the state will encounter difficulties in allocating funds for research and innovation from the private sector into the national research and innovation system. However, with a package of facilities and incentives for the relevant investors (who are directing the industry), this path can be corrected. Here are some incentives and facilities that could be provided by the state in exchange for pumping these funds into the national research and innovation system:

- Ensure intellectual property protections for the patented innovations, which extend up to twenty years, according to the current national laws and regulations. Currently, there is no confidence among investors and companies in the effectiveness of implementing intellectual property protection systems, despite the existence of laws and regulations. The creation of courts specialized in intellectual property, who resolve related disputes and consider applications for infringements, is crucial to help investors think seriously and practically about investing in patent-protected products and services.

- Issue a system as protection from external competition for at least five years in the event that a company wants to take the initiative and invest in a new business idea that does not have intellectual property protections, because risk is very high in successful new innovations and businesses. When such enterprises succeed, competition increases, and many enter this field, which wastes profits for the initiator who made the first sacrifices to enter into this field. For clarification, not every new business idea can be protected by the intellectual property system. Accordingly, we must give investors who risk their capital on business ideas that cannot be protected by the intellectual property system some protection, even if temporary.

- Many successful industries inside the country that produce "Made in Saudi Arabia" products are not national innovations. Rather, they are fully imported technologies with their production lines installed inside the country to create a national industry without the interference of any national innovations or development. Accordingly, it is necessary to create a new market based on national innovation—"Saudi innovation"— along with "Made in X". For example, there may be products made abroad in China or India, but they are Saudi innovations. This would encourage investment in intellectual property, whether manufactured at home or abroad. For example, Apple smartphones indicate "Designed in California and Manufactured in China".

- Establish a fund to support projects that will bear the "Innovated in Saudi Arabia" label. This fund will provide multiple programs to serve investors.

- Hold exhibitions more than once a year in different cities within the Kingdom by the Ministry of Investment and Trade and Chambers of Commerce to encourage products bearing this mark and provide a market for selling "Innovated in Saudi Arabia" products.

- Establish a TV program and broad marketing campaign for the brand "Innovated in Saudi Arabia".

- Establish a department in the Ministry of Investment and Trade that authorizes the development of the brand "Invented in Saudi Arabia».

- Establish a committee for scientific research chairs in charge of organizing research chair affairs in all academic sectors and guaranteeing investor rights when pumping money into these research chairs, such as a percentage of the ownership of granted patents according to the liquidity provided, as well as a distribution of royalties in the case of profits in fixed proportions. This process should also include applications of the appropriate mechanisms (referred to in a previous chapter) to prevent the leakage of funds abroad. Requests for the creation of research chairs could be determined by chambers of commerce based on national market needs.

- Exemptions from value-added taxes on all materials and services involved in the production of the innovative product or service "Innovated in Saudi Arabia" for the first five years and an exemption from half the taxes for the next five years.

- Exemptions from value-added taxes on selling an innovative product or service that was "Innovated in Saudi Arabia".

- Publish a journal bearing the names of enterprising investors who have created innovative companies or adopted/developed an innovative product from the national research and innovation system.

- Organize prizes and competitions to stimulate the transfer of innovation to the market under the auspices of the Ministry of Commerce.

Chapter 26

Developing an Index To Measure
The Contributions of Entities
and Researchers

In an attempt to solve this dilemma, we previously developed an index that contains equations that enable the decision-maker to determine the actual contributions (roughly) of the researcher and the research entity after stripping foreign contributions from the work, thus allowing the decision-maker to determine where to direct and harness resources. We have explained in detail in previous chapters the research sub-contracting system practiced by some researchers and research entities. The establishment of such an index is extremely important to determine the truth and to identify the national capabilities of research entities and researchers.

The importance of such an index increases when we want to allocate large sums of money to research chairs and trust the research chair to a distinguished Saudi scientist. How can we locate a research in his field without such an index? In a previous work, we suggested using this indicator that the impact factor of a scientific journal is not a reflection of the impact of the researcher nor of the quality of his published work. The h-index, a famous index, bears many shortcomings in omitting the total numbers of quotes of a researcher's work and does not count the researcher's contributions to or roles in published works.

Chapter 27

Oversight Board for The Credibility Of Scientific Output

It is no secret what the newspapers publish about the productions of some national research entities for products including cars, planes, tanks, space shuttles, smart phones, and others. Some of these entities apologized that the advertised products were not actually produced, which embarrassed everyone. On the whole, even if these entities do not recognize their shortcomings, where are these products in the local and international markets? Are they really there? The bottom line is that if a solution to this dilemma is not reached, then the reputation of all research entities will suffer, even if they take honest steps toward success and are close to producing products that will dazzle the world. We repeat that there is no limit to our ability to develop and innovate, but why do not we focus on honest entities?

Accordingly, we propose the creation of a monitoring body to determine the credibility of the outputs of research entities. This body should consist of experts in the fields of various technologies to review any product claimed by a research entity, determine its credibility in producing that product, and elucidate whether it is really a product of that entity or imported from abroad. Was the product made locally in the system's workshops and experimental fields, or was it imported from abroad? Is innovation the product of individuals affiliated with the national system or under the umbrella of cooperation? This body should then communicate with the press and the Ministry of Media to explain in a press conferences the credibility of what has been produced and what we should be proud of.

This is a sound start for the advancement of the national research and innovation system. We will lose our reputation and credibility by neglecting untrue allegations. We will mention here some tricks employed by research entities to falsely claim they have invented or developed a product:

- Signing a partnership contract with a distinguished foreign entity for the production of a product intended to be developed, and development process takes place at the location in which the contract is signed. To cover this process, researchers from the national entity are named in the published work even if only by making several visits to the foreign party. This makes it difficult for analysts to understand where the product was developed and innovated.

- Establishing a joint research center with a reputable foreign authority in a field of research. This center does not contain one full-time researcher, and, through this cooperation, innovations are produced and developed on the foreign side, and funds are transferred from the national side. It is thus difficult to distinguish who produced what.

- Engaging distinguished foreign researchers in the target field through several methods, including employing them as consultants or associate researchers involved in research projects or consulting contracts. Ultimately, the work of innovation or development is produced by the foreigner in his country, and, through this legal cover, the business is leased to the entity and the researchers affiliated with this entity.

There are many ways and means that such entities blind decision-makers by deluding them that the products invented or developed are the work of the national research entity. However, this involves the process of transferring funds to obtain a product by finding a legal cover for that process. Unfortunately, some national research and innovation entities have succeeded in deluding the public and the media about the quality of their work and employees. Therefore, an oversight body to determine the credibility of the outputs of research entities is imperative under our current circumstances.

Chapter 28

Council of Scientific Integrity and The Penal Sanctions Package

Being tolerant with the negligent will not rectify the current situation; warding off harm always precedes pleasure, and a lack of punishment is better than non-deterrent punishment. Some lax penalties encourage the recurrence of offenses for those who incurred the penalty first and then for others who viewed the softness of the penalty. A lack of respect for and transgression of laws is a violation of the rights of innocent people. These laws and regulations are in place to protect individual and public rights.

A person, a researcher in our case, may complain and demand redress when someone infringes upon his rights, but who serves justice for the infringement of public rights and who seeks these rights? They are statesmen. We do not pretend here that public rights are superior to private rights, but public rights are the rights of all, the wealth of all, and the reputation of a nation. It is our duty to preserve these rights because the future of all depends on everyone and not just an individual. Here, the harm is generalized, and punishment should be obligatory in the best interest of everyone.

When someone infringes upon the rights of a researcher and steals his research (or intellectual property) and attributes it to himself (or forfeits his rights and does not mention his contributions) and the wronged party finds out this transgression, he should turn to governmental bodies to demand redress. If this governmental body does not have deterrent sanctions and strict laws, and if this body is not run by competent and qualified individuals through a series of procedures put in place to enable and enforce these laws and penalties, then the aggrieved party will not benefit from his complaint, thereby allowing abuse and encroachment to continue.

When a person neglects the rights of a governmental research entity to which he belongs through affiliation to another entity when publishing his research works, who does justice to that entity? Who is permitted to waive the rights of the aggrieved government agency? It is not a gift granted by a senior official for this entity to forfeit the rights of the entity. Nor should we overlook the right of all, which is a public right.

Through my years of work in the research and innovation sectors, I have determined the three pillars upon which scientific research integrity in this country needs re-governance. These pillars are: 1) legislations, 2) penalties, and 3) the mechanisms and procedures for implementing laws. A deficiency in any of these three pillars renders the other two pillars worthless and ineffective.

In brief, these three pillars must be refined and re-installed to ensure scientific accuracy and trust in the national research and innovation system. We will lose our reputation if we do not fix this.

- The current laws, penalties, and sanctions are neither sufficient deterrents nor comprehensive enough for all types of violations. First, all possible violations should be identified by examining past violations. Then, we can develop all the penalties and sanctions that can be applied to researchers and research entities, each violation will be determined, and the punishment or penalty will be implemented according to degree of the violation.

- It is necessary to put people in power to implement this order accurately and without complaints or favoritism, when there are any such violations. These people should be free from research work to ensure no conflicts of interest. They should also be well-trained in all types of violations, and the ways in which researchers and entities circumvent violations.

• Establish a series of procedures and mechanisms to enable authorized persons to enforce the system accurately and without interference from others.

In order for these three pillars to remain upright, they must not be affiliated with any entity within the national research and innovation system to ensure no conflict of interest and to cut off any means to influence decision-making. Establishing an independent national body (or an independent national council) to control scientific integrity is essential. The newly-established Intellectual Property Authority does not control the procedures of scientific integrity for researchers, as its procedures and laws include only copyright.

Chapter 29

Unified National Scientific Council: Cancellation of scientific councils in universities

Scientific councils in universities and national research and innovation entities face three major challenges:

1. compliments and the inability to control the influence of some researchers with connections and those in high positions within the entity;

2. the asymmetry of procedures from one entity to another generates disparity in quality in recruitment and scientific promotions; and

3. the need for a full-time staff to study the applications of scientific councils, which wastes a

portion of public wealth through the abundance of these councils and their committees.

It is suggested that we establish a unified council that handles all appointment applications, academic promotions, and other researchers' requests for secondment, sabbatical leave, and others. This will save expenditures in the operation of scientific councils, relieve university rectors and their deputies from many burdens, limit the influence of those with personal connections to promotions and appointments, and prevent the large disparity in quality between university professors and researchers from one research entity to another. In this way, there will be minimum standards for all our research entities.

After the digital revolution, all applications can now be electronically submitted without direct contact with the applicant or the entity to which the applicant belongs, which increases justice and fairness in procedures, allows greater transparency in decision-making, and reduces expenses.

Chapter 30

Reforming the Financial System of the National Research And Innovation System

What is the problem with the financial system in the current national research and innovation system? What are the shortcomings to be addressed? The problems of this system are limited to two elements: the flow of funds to any governmental research entity and the spending mechanisms within the government research entity. In short, there is a challenge in introducing financial resources into an entity and a greater challenge in getting them out of the entity.

In order to accommodate a portion of the complexities of this issue, a governmental research entity is ultimately a government management body subject to the same rules applicable to all government entities in terms of defining and allocating budgets, disbursement mechanisms, contracting, and procurement. The crux of the matter lies in the two challenges. Each government agency has "budget" allocations from financial resources at the beginning of each year, regardless of what it requires (an increase or a decrease). Then, the entity spends its funds based on the restrictions and regulations referred to in the "Procurement and Competition Law". The entity should return all undisbursed amounts to the Ministry of Finance at the end of each year. The irregular flow of funds on an annual basis is a challenge, which makes medium-term plans difficult to achieve due to the unpredictability of the amount of liquidity each year.

For the government agencies that obtain income such as fees or taxes, they return such income to the Ministry of Finance at the end of each year and spend it on their expenses based on the budget allocated to them for this year, regardless of what they receive. Accordingly, any income received by the research entity is not supposed to be spent within the entity but rather returned to the Ministry of Finance. Although some of these systems have been improved by giving universities freedom to expend the income they receive, this process is not fully developed polished and still contains gaps, in addition to the variations in procedures from one research entity to another.

What we have mentioned above is an introduction to help facilitate government research and innovation entities to receive cash from the private sector for carrying out research, product development, or innovation—for example, manufacturing a stereo, improving a computer program, or transferring knowledge or technology. Yes, there are many cases where funds have been transferred from the private sector to the academic sector, but there are no clear structural management frameworks that determine the entities' authority to receive these funds or how to dispose of them. All the cases examined are governed by individual contracts, each of which may not be identical with the other; moreover, some of their contents contradict the governmental competition and procurement system at the time of disbursement and spending.

- On this basis, we offer five basic proposals to reform the financial system in the national research and innovation system:

- Establish a unified regulation to enable government research and innovation entities to market their products, attract funds from the private sector, and establish financial accounts for that purpose.

- Do not apply the system of government procurement and competition to the attracted funds from the private sector; the competition system should apply only to the general operating budget of the entity.

- Establish accounts in commercial banks separate from the accounts of the governmental research entity to establish a separate account for each investor in the university with a project or initiative.

- Establish an account in commercial banks for every researcher who receives financial support under the supervision of a governmental research entity. If the research entity has a thousand researchers, then there is nothing wrong with creating a thousand accounts. This enables the research entity to track and control each account in a manner that does not prejudice the quick and effective use of the account.

- Create a general directory for the researchers and investors in each national research and innovation entity responsible for creating account books, recording and reviewing entries, and settling accounts for all sub-accounts. This would serve as a directory whose functions are different from the main financial department of the entity, as it is devoted to the funds received (and spent) outside the annual budget allocated to the entity.

For further clarification of the above, in the normal cases where these accounts (which we recommend creating) do not exist, any investing party (a party that desires to develop a product or innovation) can deposit cash money to the main entity account, which is a management and financial error. The government entity is supposed to receive taxes and fees and return them to the state budget by the end of the year. On the other hand, another very common mistake is made by depositing some cash into accounts in commercial banks. These accounts are for personal and individual use and not for governmental research entities. Therefore, an entity has no right to request account statements, track restrictions, or make a settlement to these accounts, in addition to the possibility of the researcher's death. In this case, the entity would face problems with the researcher's heirs. If this is a separate account in a commercial bank that branches out from the primary entity's account, then the entity has all power to track the account, recover the amounts in cases of necessity, and transfer authority to dispose of the account to another new or participant researcher. This would empower governmental research entities and strengthen financial governance.

CONCLUSION

I did not want to publish this work, as this is a summary of my ideas and insights generated over the past decade and a half, and I was trying to preserve them for myself. I intended to reflect upon these concepts and apply them to the national research and innovation system through my presence as a leader and manager in one of these national entities. After my attempts to implement such changes, I learned that it is necessary to formulate these ideas and opinions for my colleagues so that we can work together with a common goal and clear concepts.

If I am being critical here, then I am criticizing myself first for being a part of this national system, in whose orbit I still run, in whose space I still swim, and in whose domains I still wander. Here, I have presented solutions and perceptions that would enable us to overcome our stumbling blocks, bridge our faults, and correct the aforementioned problems.

I am sure that things will be straightened out after a while. The national research and innovation system for this country is a model that can be repeated or expanded after its success to include a larger system comprising the research and innovation system of the Arab and Islamic nations, the system that once dazzled humanity and was a cause of the Earth's prosperity.

Mohammed Al-Shamsi

ISBN 978-1-7346287-0-8 (Printed Version)

ISBN 978-1-7346287-7-7 (Electronic Version)

$24.00
ISBN 978-1-7346287-0-8
52400>

9 781734 628708

www.ingramcontent.com/pod-product-compliance
Lightning Source LLC
Chambersburg PA
CBHW031958190326
41520CB00007B/293